Planet Earth
NOBLE METALS

TIME® LIFE BOOKS

Other Publications:

AMERICAN COUNTRY
VOYAGE THROUGH THE UNIVERSE
THE THIRD REICH
THE TIME-LIFE GARDENER'S GUIDE
MYSTERIES OF THE UNKNOWN
TIME FRAME
FIX IT YOURSELF
FITNESS, HEALTH & NUTRITION
SUCCESSFUL PARENTING
HEALTHY HOME COOKING
UNDERSTANDING COMPUTERS
LIBRARY OF NATIONS
THE ENCHANTED WORLD
THE KODAK LIBRARY OF CREATIVE PHOTOGRAPHY
GREAT MEALS IN MINUTES
THE CIVIL WAR
COLLECTOR'S LIBRARY OF THE CIVIL WAR
THE EPIC OF FLIGHT
THE GOOD COOK
WORLD WAR II
HOME REPAIR AND IMPROVEMENT
THE OLD WEST

For information on and a full description of any of
the Time-Life Books series listed above, please call
1-800-621-7026 or write:
   Reader Information
   Time-Life Customer Service
   P.O. Box C-32068
   Richmond, Virginia 23261-2068

This volume is one of a series that examines the
workings of the planet earth, from the geological
wonders of its continents to the marvels of its
atmosphere and its ocean depths.

Cover
One of the most spectacular examples of natural
gold leaf ever found, this seven-inch-high, one-
pound specimen comes from California's Eureka
mine. Its shape was dictated by the rocky fissure
in which it formed.

Planet Earth

NOBLE METALS

By Jeffrey St. John
and The Editors of Time-Life Books

Time-Life Books, Alexandria, Virginia

Time-Life Books Inc.
is a wholly owned subsidiary of

TIME INCORPORATED

FOUNDER: Henry R. Luce 1898-1967

Editor-in-Chief: Henry Anatole Grunwald
Chairman and Chief Executive Officer: J. Richard Munro
President and Chief Operating Officer: N. J. Nicholas Jr.
Chairman of the Executive Committee: Ralph P. Davidson
Corporate Editor: Ray Cave
Executive Vice President, Books: Kelso F. Sutton
Vice President, Books: George Artandi

TIME-LIFE BOOKS INC.

EDITOR: George Constable
Executive Editor: Ellen Phillips
Director of Design: Louis Klein
Director of Editorial Resources: Phyllis K. Wise
Editorial Board: Russell B. Adams Jr., Dale M.
Brown, Roberta Conlan, Thomas H. Flaherty, Lee
Hassig, Donia Ann Steele, Rosalind Stubenberg, Kit
van Tulleken, Henry Woodhead
Director of Photography and Research:
John Conrad Weiser

PRESIDENT: Christopher T. Linen
Chief Operating Officer: John M. Fahey Jr.
Senior Vice Presidents: James L. Mercer,
Leopoldo Toralballa
Vice Presidents: Stephen L. Bair, Ralph J. Cuomo, Neal
Goff, Stephen L. Goldstein, Juanita T. James, Hallett
Johnson III, Carol Kaplan, Susan J. Maruyama,
Robert H. Smith, Paul R. Stewart, Joseph J. Ward
Director of Production Services: Robert J. Passantino

PLANET EARTH

EDITOR: Thomas A. Lewis
Deputy Editor: Russell B. Adams Jr.
Designer: Albert Sherman
Chief Researcher: Patti H. Cass
Editorial Staff for *Noble Metals*
Associate Editor: Neil Kagan (pictures)
Text Editors: Sarah Brash, Jan Leslie Cook,
Thomas H. Flaherty Jr.
Researchers: Janet Doughty, Gregory A. McGruder
(principals), Barbara Brownell, Roxie M. France,
Barbara Moir, Judith W. Shanks
Assistant Designer: Cynthia T. Richardson
Copy Coordinator: Elizabeth Graham
Picture Coordinator: Renée DeSandies
Editorial Assistants: Caroline A. Boubin, Mary Kosak

Special Contributor: Karen Jensen (text)

Editorial Operations
Copy Chief: Diane Ullius
Editorial Operations Manager: Caroline A. Boubin
Production: Celia Beattie
Quality Control: James J. Cox (director)
Library: Louise D. Forstall

Correspondents: Elisabeth Kraemer-Singh (Bonn);
Maria Vincenza Aloisi (Paris); Ann Natanson
(Rome). Valuable assistance was also provided by:
Mirka Gondicas (Athens); Brigid Grauman
(Brussels); Robert Kroon (Geneva); Peter
Hawthorne (Johannesburg); Lesley Coleman,
Millicent Trowbridge (London); Christina Lieberman
(New York); Michael Kepp, Gavin Scott (Rio de
Janeiro); Anne Wise (Rome).

TIME-LIFE is a trademark of Time Incorporated U.S.A.

Library of Congress Cataloguing in Publication Data
St. John, Jeffrey.
  Noble metals.
  (Planet earth)
  Bibliography: p.
  Includes index.
  1. Precious metals. I. Time-Life Books. II. Title.
III. Series.
TN759.S7 1984   669'.2   83-15923
ISBN 0-8094-4504-2
ISBN 0-8094-4505-0 (lib. bdg.)

THE AUTHOR
Jeffrey St. John is a writer and commentator based in Washington, D.C. He is the author of several monographs and books on national and international affairs; he writes the weekly syndicated feature "Headlines & History" and appears regularly on various radio and television public affairs programs.

THE CONSULTANTS
Dr. Peter C. Keller is Director of Education at the Gemological Institute of America in Santa Monica, California. He is former Curator of Mineralogy at the Los Angeles County Museum of Natural History and has published numerous articles on gold.

Donald Chaput, formerly research director of the Michigan Historical Commission, has been curator of history at the Natural History Museum of Los Angeles since 1972. He is primarily interested in the mining and milling history of the Americas, Australasia and the Philippines. His book, *The Cliff: America's First Great Copper Mine,* was published in 1971.

J. Roger Loebenstein is a physical scientist with the Bureau of Mines, United States Department of the Interior. He has worked as the commodity specialist on platinum-group metals and is a member of the International Precious Metals Institute.

CONTENTS

CREATIONS OF NATURE'S FORGE

Noble metals are most familiar in highly refined and polished forms — jewelry, coins and glorious works of art. Yet some of the rarest and most beautiful silver and gold objects on earth owe none of their splendor to human craft.

Although they are typically found as rounded nuggets or in rockbound veins, gold and silver are by nature crystalline; if the conditions of their formation permit, they will assume gemstone-like shapes of geometric symmetry. But the ideal conditions of pressure and temperature that allow the minerals' atoms to assume their characteristic arrangements on a large scale are rare. Moreover, metallic crystals must have plenty of room in which to form; the best crystals develop in spacious rock-lined pockets where their growth is not inhibited.

In most cases, the nascent metals are forced to compromise for lack of space.

Gold formations sometimes exhibit flattened triangles called trigons, which are truncated crystals that ran out of the room they needed to achieve their third dimension. But even this evidence of the mathematically ideal potential of gold and silver is extremely rare. Ruled by the geologic whims of the surrounding rock, the metals are pressed into natural sculptures of a diversity and intricacy seldom surpassed by human artistry.

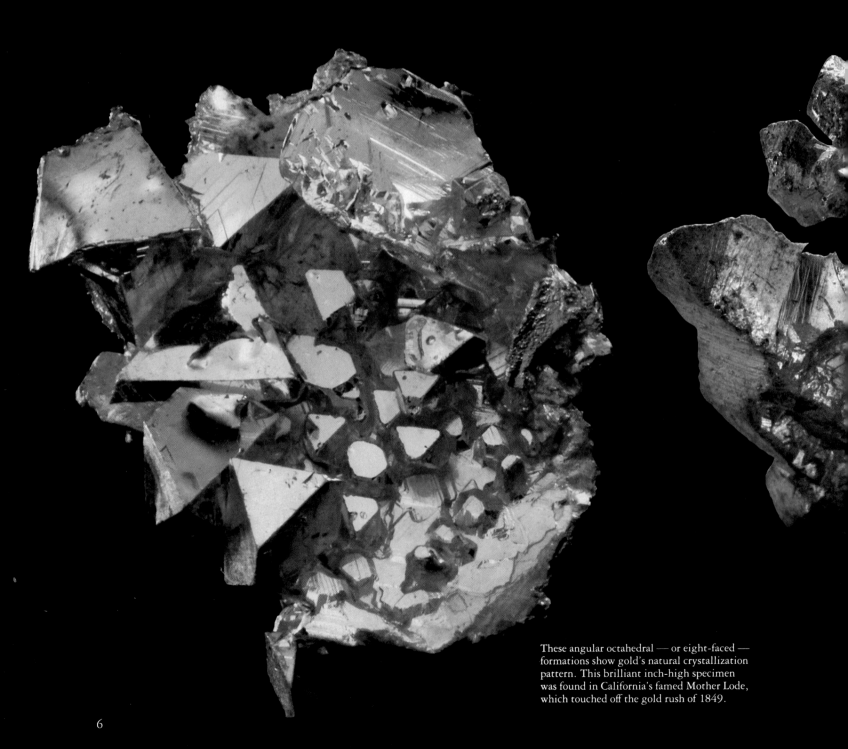

These angular octahedral — or eight-faced — formations show gold's natural crystallization pattern. This brilliant inch-high specimen was found in California's famed Mother Lode, which touched off the gold rush of 1849.

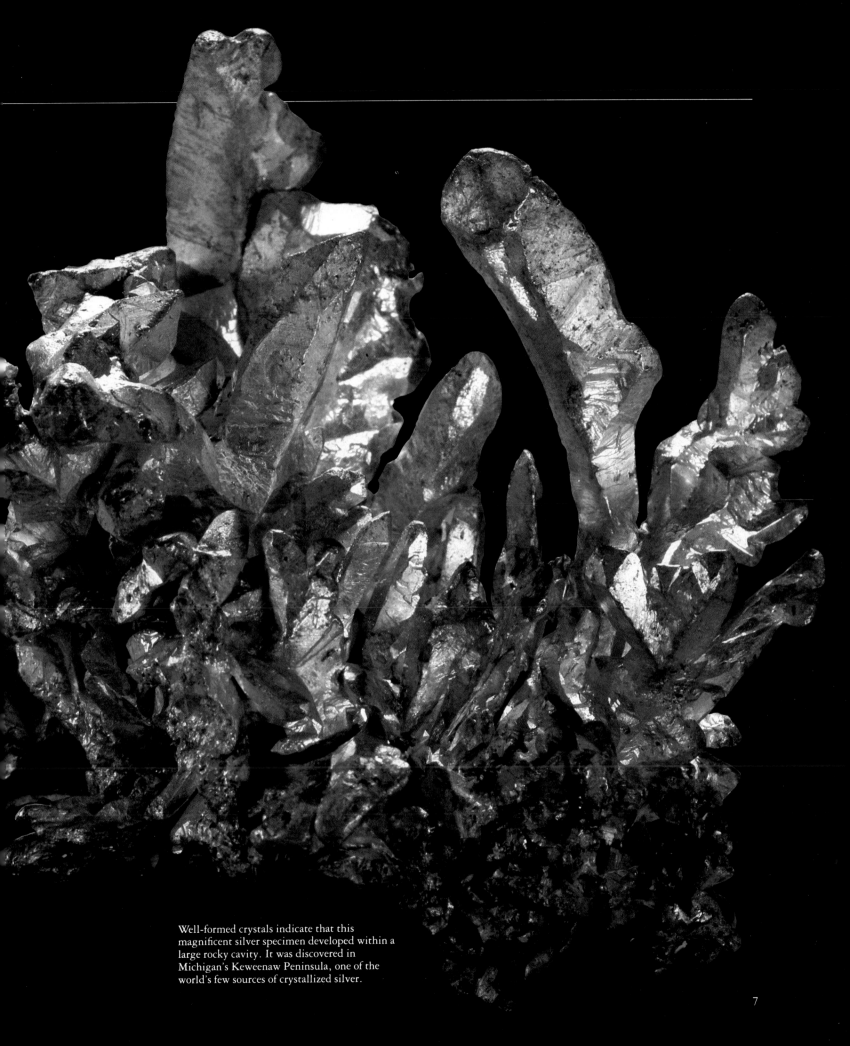

Well-formed crystals indicate that this
magnificent silver specimen developed within a
large rocky cavity. It was discovered in
Michigan's Keweenaw Peninsula, one of the
world's few sources of crystallized silver.

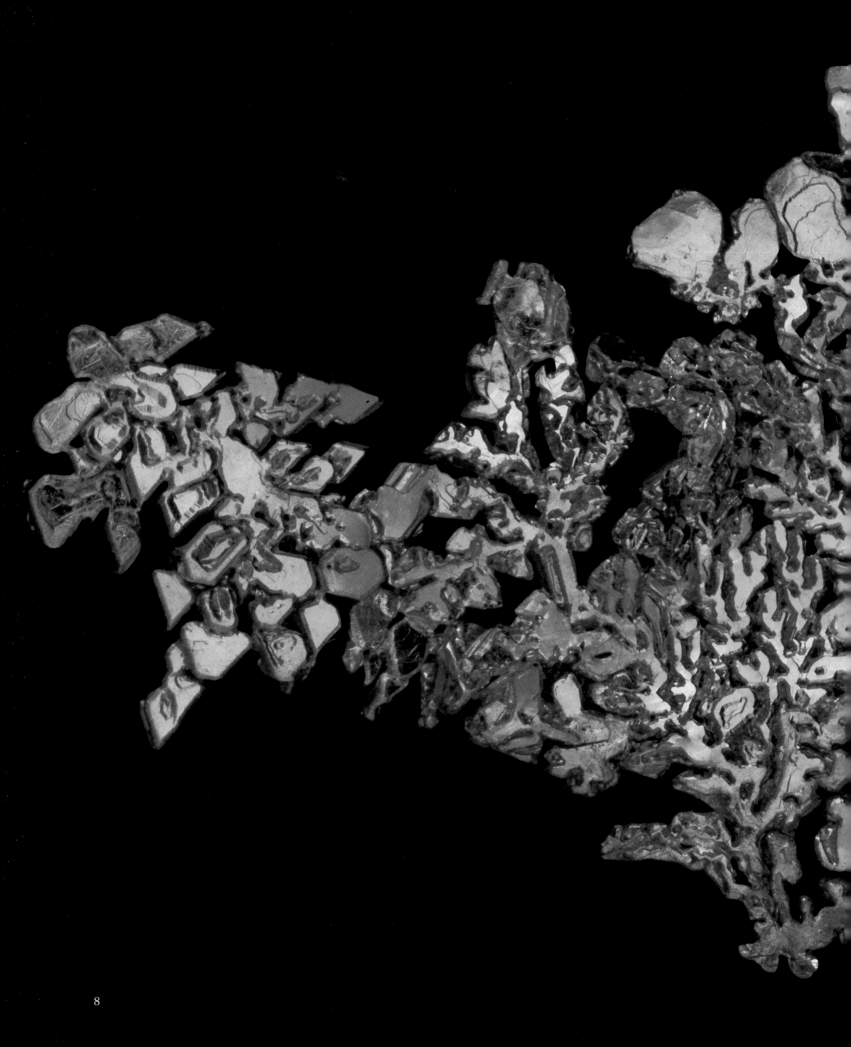

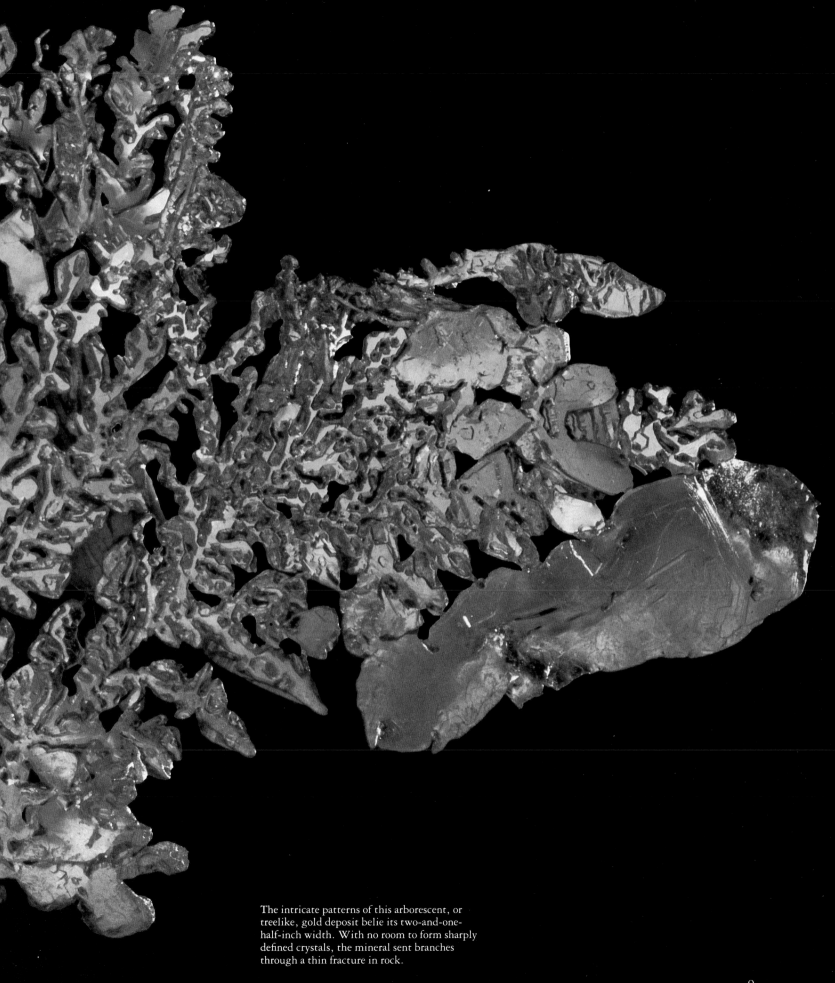

The intricate patterns of this arborescent, or treelike, gold deposit belie its two-and-one-half-inch width. With no room to form sharply defined crystals, the mineral sent branches through a thin fracture in rock.

9

This thin sheet of silver, about five inches
high, bears surface striations that were etched by
movements of its host rock. The specimen
took shape in a fractured rock, where constricted
space caused flattened crystals.

Sprinkled with tiny triangular trigons —
flattened crystal faces — leaf gold gilds the
quartz crystals on which it developed. Pockets
within quartz veins are the most common
location of crystallized gold.

11

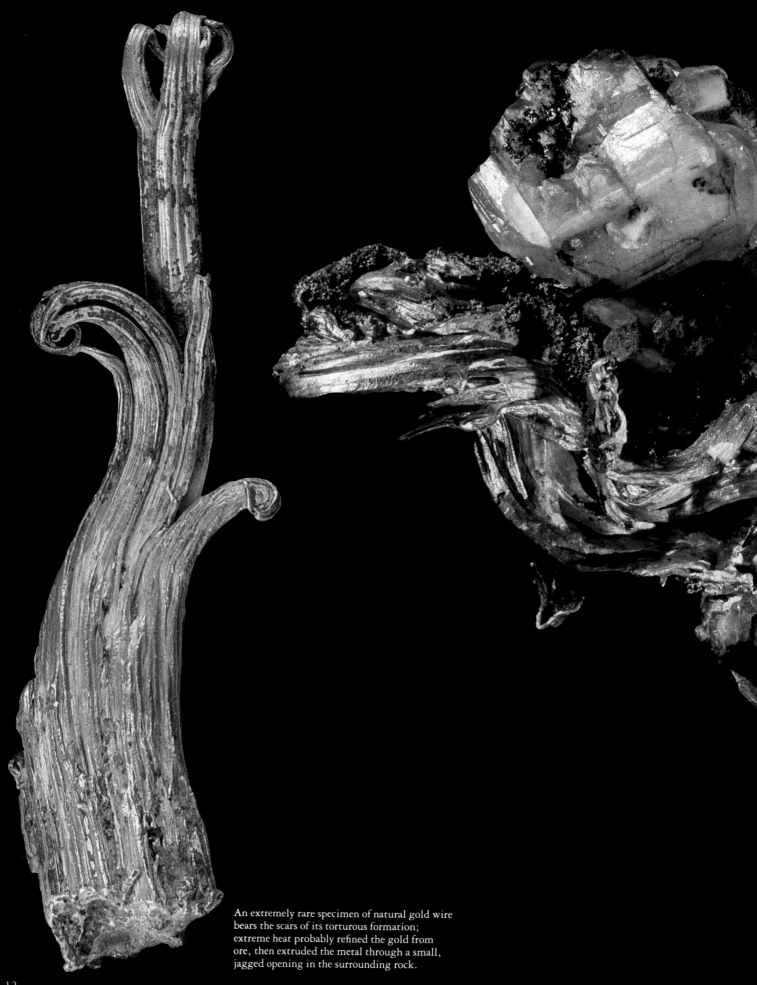

An extremely rare specimen of natural gold wire
bears the scars of its torturous formation;
extreme heat probably refined the gold from
ore, then extruded the metal through a small,
jagged opening in the surrounding rock.

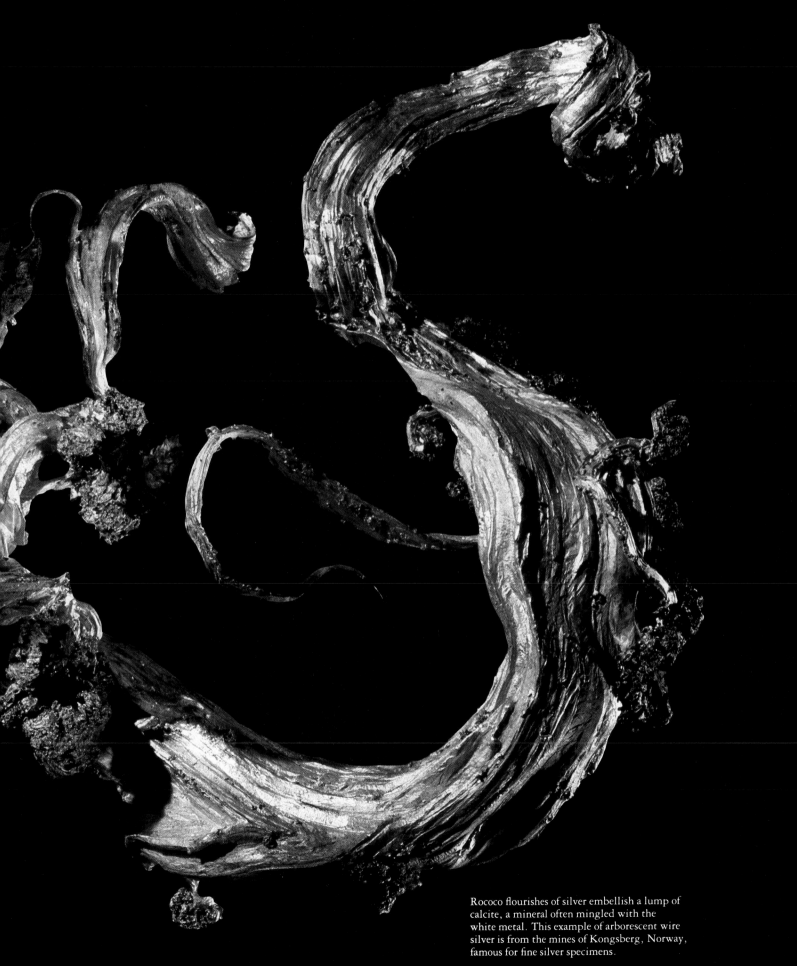

Rococo flourishes of silver embellish a lump of calcite, a mineral often mingled with the white metal. This example of arborescent wire silver is from the mines of Kongsberg, Norway, famous for fine silver specimens.

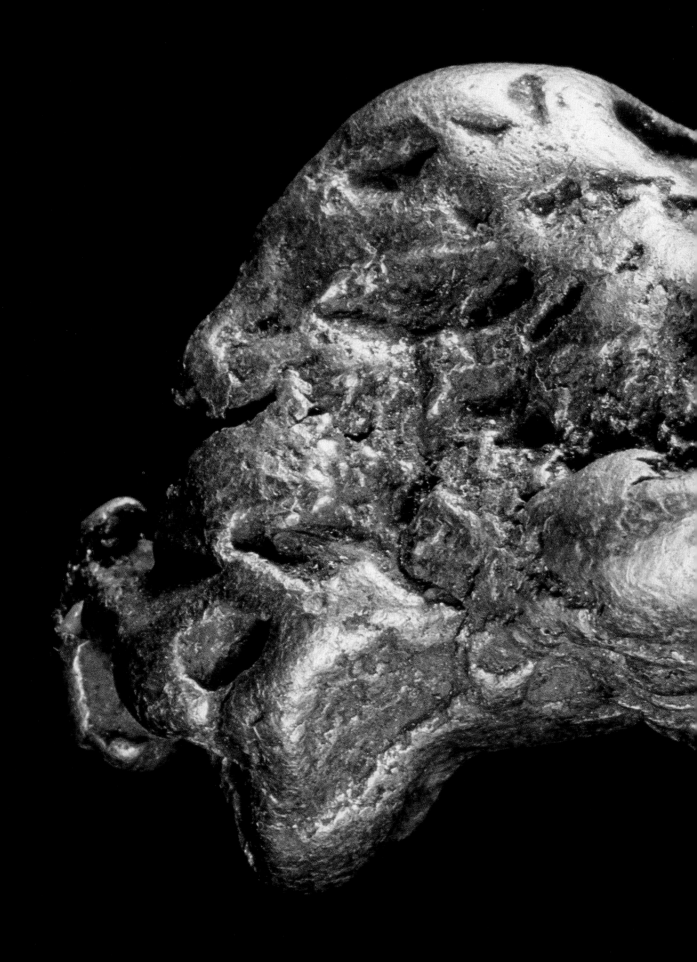

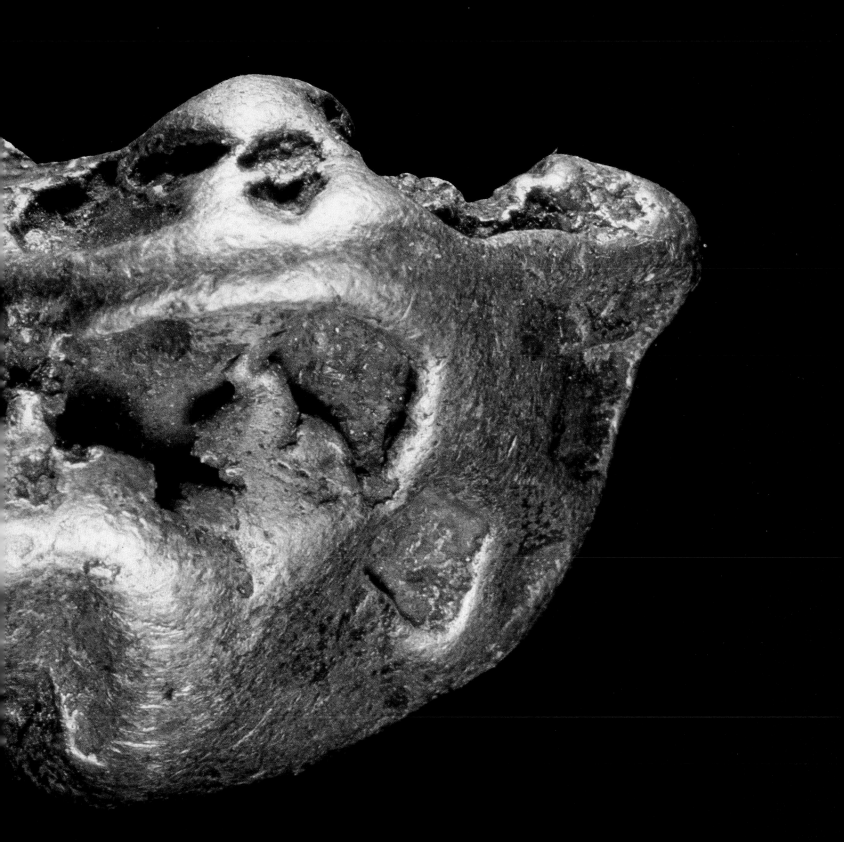

Although a gold deposit may begin as a crisp array of crystals, this nugget represents the form in which gold is usually discovered. Shorn from its rocky home by erosion, bounced down hillsides and washed through rivers, the highly malleable metal's sharp edges were quickly beaten into rounded bumps.

Chapter 1

AN ETERNAL QUEST FOR RICHES

The Sumerian King was dead. In the city of Ur, his capital, the royal court marched solemnly behind the King's bier to the burial site. The sun's rays created a glittering dance of light as they played across the gold and silver artifacts that adorned the monarch's body. The procession descended into a deep, sloping pit to the accompaniment of a rhythmic musical sound coming from the gold and silver jewelry worn by members of the court. Some of the courtiers carried more useful objects: Military officers bore golden weapons; musicians held lyres and flutes made of silver. The King's favorite chariot, worked with both silver and gold, was drawn into the burial pit by oxen. Three score ladies of the court took their places, wearing intricate gold and silver headdresses befitting their rank.

When the procession ended and the priest's prayers were done, the court members drank deeply from gold and silver cups. In a few minutes the narcotic brew had done its work; the courtiers fell into a deep and final sleep. Other citizens of Ur, arrayed at the mouth of the pit, spread a blanket of earth over the royal bier and the lifeless attendants. When the pit was full, a shrine was erected above it to mark the spot where the entire Sumerian court had made the ultimate demonstration of fealty to their sovereign.

Ur was a city of ancient Mesopotamia, located inland from the Gulf of Persia in what is now southern Iraq. The funeral ceremony is now known to have taken place early in the third millennium before the birth of Christ. Until relatively recently, however, historians believed that the Sumerian civilization of this period existed only in myth.

If not for the tenacious permanence of gold and silver, the highly developed culture of ancient Ur might have remained unknown, its glories lost forever to the ravages of time. Then, in 1927, the British archeologist Sir Leonard Woolley uncovered the first of 16 royal tombs at Ur. In the final stages of excavating the burial site, Sir Leonard and his wife sent away their team of diggers so that they could proceed undisturbed, at a careful pace. For 10 days, spent mostly lying on their stomachs from dawn to dusk, the couple painstakingly used knives, brushes and even their breath to clear away the last of the dirt and debris. The skeletal remains they uncovered had turned almost to dust, but the treasure trove of jewelry, weapons, utensils and other objects made of gold and silver remained uncorrupted by time or nature. With the help of an inscribed foundation stone, they could be dated to roughly 2700 B.C. and thus helped to establish the First Dynasty of Ur as historical fact rather than mere legend. Moreover, the artistic designs showed that early Sumerian metalsmiths had achieved a sophistica-

A stern-visaged Mycenaean burial mask, crafted of cold-hammered gold around 1550 B.C., was thought by its discoverer to be the mask of Agamemnon, legendary leader of the Trojan War. But later research showed that the monarch represented here died some 300 years before Agamemnon sailed for Troy.

17

tion that was believed to have developed only much later—and not in Sumeria but in ancient Egypt.

Great durability is a primary quality that sets apart gold and silver—and the more recently identified platinum family—from the other 100 or so elements that constitute all earthly matter. Gold, silver and platinum are called the noble metals, a name that in chemical terminology refers to their outstanding resistance to the corrosion and oxidation that cause base metals such as iron, copper and tin to weaken and disintegrate. But permanence is only one of many attributes that through the ages have earned gold, silver and platinum the accolade "noble" in a broader sense. The metals are strong and heavy, but they are at the same time wonderfully malleable. They can be drawn out to form extremely thin wire and thread, or forcibly shaped without fear of crushing or shattering. They are pleasing to the touch and delightful to the eye. And though the noble metals may be found throughout the world, they are rare enough to stimulate the acquisitive drive in human nature.

Of the three, gold is supreme: unmatched for versatility, durability and the timeless fascination of its lustrous yellow sheen. It is said that the history of gold is the history of the world. Portable and universally accepted, gold became the coinage of trade between peoples and nations, the motivator of conquering armies and voyages of exploration. It is the stuff on which empires have been built, and for lack of which they have fallen. No civilization has risen to greatness without it.

Generations of pseudo scientists called alchemists strove to create gold from lead and other base metals. They labored in vain—and failure sometimes cost them their heads. But their efforts paved the way for the modern sciences of chemistry and physics. Today, gold's nobility makes it ideal for many uses in industry, and for special applications in the exploration of outer space.

Though men and women have searched for gold throughout history, the sum total of their harvest is not large: about 100,000 tons. Because gold endures, however, most of that harvest still exists. One civilization digs up what earlier ones have buried—usually to bury it again. Thus some of the gold in the United States reserve at Fort Knox, Kentucky, may once have been the prized possession of people as far back as the late Stone Age.

Gold is one of the few metals found in nature in a free state, and one can imagine members of a hunting party 40,000 years ago being attracted by the shimmering specks and nuggets lying at the bottom of a shallow stream. The living beauty of the metal must have filled the hunters with awe. Certainly it was a striking contrast to the drabness of their daily lives. Brute existence in dark caves lighted only by campfires afforded them little esthetic pleasure.

Yet here was an object that glittered intriguingly both in sunlight and by firelight. It was almost indestructible and yet it could be cut by flint. Even when cold, it could be hammered into new shapes. When heated in a fire it became even more malleable and could be worked into intricate designs.

For people of the earliest civilizations, gold was the metal of the sun—a natural connection that has persisted through the centuries. (The Hebrew word for the gleaming substance was *zahev*, meaning "illumined by the sun's rays." Ancient Inca called particles of gold "the sweat of the sun.")

An elaborate golden headdress encircles the skull of a female member of the royal court at Ur, one of scores of courtiers entombed with their King around 2700 B.C. Although crushed by the weight of the earth, the hammered-gold headdress remained in such good condition that it could have been reassembled; this one was preserved just as it was unearthed in 1927.

And since the sun was the first of the deities, many of the initial uses of gold were religious. Simple cold-hammered amulets of gold have been found in Spanish caves used by Stone Age families as places of worship. The Babylonians, successors to the Sumerians in Mesopotamia, raised soaring pyramids, shaped like tiered wedding cakes, to their sun god. They sheathed the highest reaches of the holy structures in gold.

The pharaohs of ancient Egypt considered themselves offspring of their sun god, and they virtually immersed themselves and their surroundings in gold, the divine metal. In the first system of hieroglyphs, the symbol for gold was a necklace. The Pharaoh Thutmose I rode forth in the 16th Century B.C. to conquer Babylon and much of Asia Minor in a chariot made of a pale gold called electrum (actually an alloy containing a variable portion of silver). The golden booty Thutmose brought home to Egypt was supplemented by a steady flow from the mines of Nubia (now the Sudan) and the Arabian peninsula and, later, from the southeast coast of Africa.

The Egyptians developed mining as an industry. Their pits and underground diggings were manned by thousands of slaves who had been captured in battle. At the command of the pharaohs, Egyptian goldsmiths created works of metallic art more exquisite than those of Sumeria. The kings and nobles of Egypt considered gold to be more than a symbol of

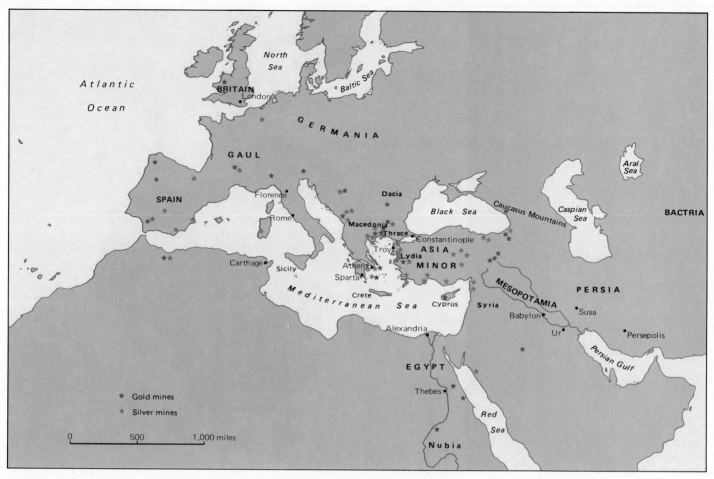

A map shows the distribution of the gold mines (*red stars*) and silver mines (*blue stars*) that provided the wealth of the cities, states and empires surrounding the Mediterranean Sea between about 3500 B.C. and 300 A.D. At the height of the Roman Empire in 117 A.D., every single gold and silver mine then known was under its domain.

power and prestige in life. It was a means of achieving immortality. Encased after death in a new body of gold, a pharaoh could make the long spiritual journey through the heavens to join his father, the sun. To help him on the journey, his tomb was filled with articles of gold, silver and precious stones.

To safeguard their mummies and their gold, the Egyptian rulers built great pyramids of stone along the Nile. But the pyramids were not enough to deter grave robbers, who invaded the sacred tombs and stripped them of their gold. For centuries it was the loot from these royal cemeteries that provided much of the gold used in trade by the peoples of the Mediterranean.

Thutmose I, seeking to foil the grave robbers, decided to conceal his grave instead of marking it with a pyramid. He had his burial chambers built in great secrecy into a limestone cliff near the city of Thebes. Succeeding pharaohs followed his example. But the robbers were persistent. Over the centuries, Thutmose's grave and most of the others were discovered and stripped of their valuables.

The thieves did not find all of the entombed riches, however. In the fall of 1922 the British archeologist Howard Carter and his patron, Lord Carnarvon, unearthed the greatest treasure trove in all of history. It was the tomb of Tutankhamen, who had lived and died — prematurely at the age of 18 — about two centuries after Thutmose I. Carter later described the moment of entering the grave, which was lighted only by the candle in his hand: "At first I could see nothing, the hot air escaping from the chamber causing the candle flame to flicker. But presently, as my eyes grew accustomed to the light, details of the room within emerged slowly from the mist: strange animals, statues and gold — everywhere the glint of gold."

A silver figurine bedecked with golden ears, breasts and boots, represents a Hittite goddess of fertility. The Hittites, who lived in Asia Minor (now Turkey and environs) from about 1900 to 1200 B.C., were known for their uncompromising devotion to their many deities. This crude but bold four-inch-high statuette is characteristic of the Hittites' early art.

20

For its journey to the afterworld, King Tutankhamen's mummified body had been encased within three coffins, one inside the other. The outer two were made of hammered gold over wooden frames; the inner one was fashioned of solid gold. The lid of the outermost coffin, seven feet long, was a golden likeness of the Boy Pharaoh, held by two winged goddesses, that, according to Carter, shone "as brilliantly as the day the coffin was made." Buried near Tutankhamen were chariots of pale gold, golden couches and idols, and statues in human form dressed in gold from their sandals up. Tutankhamen's throne was there, too, fashioned of beaten gold 20 inches thick. On its back panel were the sculpted figures of the young monarch and his queen, Ankhesenamen. From above them the rays of a golden sun reached down upon their heads, suggesting a benediction by the all-powerful sun god on a mortal daughter and son.

Egypt at its zenith enjoyed the first golden age, a descriptive term found in its own annals. But when Egypt lost its place in the sun, gold remained as a constant and monumental force, affecting the course of many of the most familiar historical events. Some scholars believe, for example, that it was not only the smile of Helen but also the magnetism of gold that launched a thousand Greek ships against Troy in the 12th Century B.C. That fortress city had been built near the Dardanelles, the narrow waterway that dominated the trade route between Europe and Asia. The Trojans grew wealthy by demanding tribute from every ship and caravan that passed. The extent of Troy's wealth was denigrated as myth until the German archeologist Heinrich Schliemann excavated the site of the city in the 1870s. Schliemann unearthed almost 9,000 gold and silver artifacts that Troy's defenders had buried for safekeeping before the Greeks plundered the city.

Croesus, King of the Lydians, was probably the first to mint gold and silver into coins, in about 550 B.C. The Egyptians had traded with small rings of gold and silver and with bars of precious metal called talents. A talent weighed about 8.5 grams and could be held in the palm of the hand. (The Chinese had been using coins for centuries but their content remains uncertain because the Chinese used the word *chin* both for gold and for metals in general.)

The riches of Croesus came from a river in his realm named the Pactolus. The origin of the river's plentiful and easily recovered store of alluvial gold was explained allegorically in the legend of Midas, an earlier King who, like his people, was desperately poor. But Midas was a man of good heart who shared what little he had. When the gods, to reward him, granted him one wish, the simple King asked that everything he touched turn into gold.

But the wish fulfilled became a curse. Every object or living thing Midas touched did turn to gold. Even food and drink hardened in his throat. Distraught, the King prayed to be rid of the cursed touch, and the gods instructed him to bathe in the waters of the River Pactolus. The river then assumed his power and its sandy bottom became for centuries a seemingly inexhaustible source of gold. The Midas legend reflected a changing attitude toward gold: It was not only a means of paying tribute to the gods and a symbol of royalty but a lure for fools as well.

Silver, more plentiful in nature than gold, generally has been a junior partner to gold throughout history. An exception was Athens—a society rooted in silver. Mined in abundance at Laurion, on the Aegean Sea, silver

The Ancient Craft of Goldbeating

More than 5,400 years ago the Egyptians discovered a spectacular application for one of gold's basic properties — its phenomenal malleability. When the precious metal was hammered into thin sheets and applied to such things as mummy cases and furniture, the lustrous gleam of small quantities of gold could be used to decorate royal surroundings in stunning profusion.

The basic methods of goldbeating and its companion craft of gilding, diagramed here, have changed little since the days of the pharaohs. Mechanized hammers now supplement, rather than replace, the exacting efforts of skilled artisans; modern machinery reduces the amount of labor involved and enables the achievement of unprecedented wispiness. A nugget of gold the size of a lump of sugar can be hammered into 108 square feet of gold leaf. The resulting sheet of gold is a mere one thousandth the thickness of this page, so delicate that a gentle breath or the touch of a fingertip would crumple it irreparably. Once used for the thrones and tombs of kings, the gold leaf of today is far more ubiquitous and adds glitter to such everyday objects as labels, book covers, chinaware and picture frames.

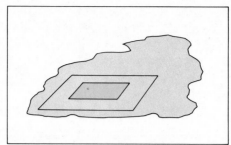

Each square of gold is placed on a sheet of goldbeater's skin — a tough membrane of animal intestine — which protects the thinning gold from the blows of the hammer.

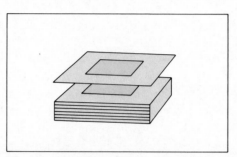

The sheets of goldbeater's skin — trimmed to four-inch squares — and the two-inch-square sections of flattened gold are interleaved in stacks of perhaps 400 of each.

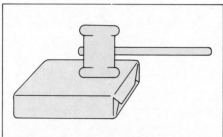

Wrapped tightly in strong paper to keep the gold and the goldbeater's skins in alignment, the entire stack is beaten until the gold sheets have doubled in size.

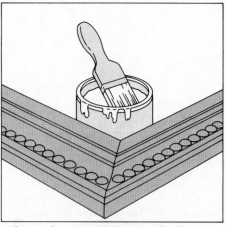

Before an object is gilded with gold leaf, its surface must be specially prepared. Here a mixture of chalk and glue, called gesso, is applied to an ornately carved wooden frame.

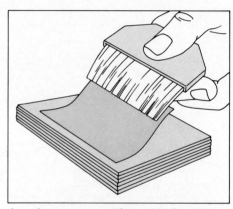

An artisan uses a special brush called a gilder's tip to lift the delicate gold leaf. A minute trace of oil, obtained by running the tip gently over the worker's hair, grips the gold.

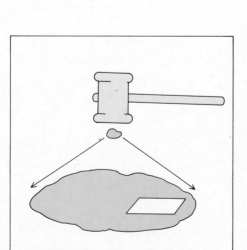

Goldbeating begins with the flattening of a measured lump or ingot of gold. The wafer is then cut into sections about two inches square for further processing.

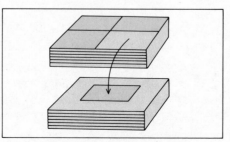

The gold is quartered, restacked and beaten until it is about $1/250,000$ of an inch thick. The original nugget of gold has yielded about 1,200 four-inch-square sheets of foil.

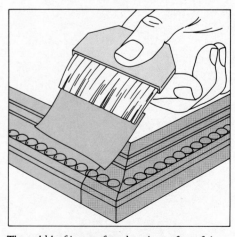

The gold leaf is transferred to the surface of the frame, gently tamped into place and then polished. The result is a wooden object that appears to be made of solid gold.

enabled the Athenians to create a uniform currency, construct a formidable navy and turn back repeated invasions by the Persians. The Athenians minted a silver coin called the drachma. They impressed each coin with the figure of an owl that symbolized the integrity of the state, much as the American eagle would be used more than 20 centuries later. The Athenian owl on a coin guaranteed the coin's weight and metal content.

What gold the Athenians had was mined in Macedonia and Thrace and was hoarded for emergencies. When Pericles rebuilt Athens after a ruinous war in the Fifth Century B.C., he commissioned a statue of the goddess Athena that would stand guard in the Parthenon over the city's gold reserve. The 38-foot statue had a face of ivory and was adorned with 2,500 pounds of gold. With fatalistic foresight, the Athenians fastened the gold to the figure of Athena in such a way that it could be quickly removed. Indeed, the statue was no more than 10 years old when an emergency arose — war with Sparta — and Athena was stripped of her gold to finance the campaign. While the city-states fought one another into exhaustion, the Persians opportunistically struck again and took over the mines that had made possible Greece's most prosperous age.

At the peak of its power, imperial Persia controlled every major source of gold from the Indus River to the Nile. Tribute in gold flowed into Susa, the Persian capital, from 30 vassal states. Even once-mighty Egypt contributed 40,000 pounds annually. King Darius I used some of his nation's wealth to build himself a magnificent new capital at Persepolis and buried much of the rest in the form of bullion.

Persian gold was a primary target of Alexander the Great's nine-year campaign of world conquest in the Fourth Century B.C. Alexander's father, King Philip II, had begun the task of empire building by recapturing the mines of Thrace and Macedonia from the Persians. Alexander, after securing Egypt and routing the army of Darius III, seized and stripped the gold-rich places in the Near East: Susa, Persepolis and finally Bactria in present-day Afghanistan. Instead of hoarding the captured gold, as the Persian kings had done, Alexander distributed it swiftly and liberally among his fighting men; even on the march, the soldiers of Alexander carried an assortment of spoils that included gold and silver bullion and coins, golden idols and other artifacts. When the survivors of his phalanxes returned home and dispersed, gold began to circulate in abundance, stimulating trade and commerce. Alexander died in 323 B.C. at the age of 33. The effect of his openhandedness on the prosperity of the West far outlasted his empire.

The Roman Empire, which succeeded Alexander's, was built primarily on silver and gold. There was little other basis for its economy; Rome had neither agricultural surpluses nor manufactured goods to trade for the commodities its people needed. The mines and accumulated treasures of distant countries were the targets of Rome's legions. Over a period of three centuries, Roman arms came to dominate all the profitable mining districts in the known world, from Spain to Britain to Dacia (modern Rumania) to Asia Minor. Roman coins, the silver denarius and the gold aureus, became the city's one true export.

Julius Caesar was drowning in debt when he was posted to Spain as governor in 61 B.C., at a time when the mines and mineral-rich rivers of Spain had already been supporting foreign masters for centuries. Rome had seized them from Carthage in the Punic Wars and extracted from them

SPLENDID TOMB OF THE BOY PHARAOH

Gold is often found in rugged mountains, but rarely in so exquisite a form as the treasure that was discovered in 1922 beneath the barren peaks of Egypt's Valley of the Kings. There, after a five-year search, archeologist Howard Carter unearthed the tomb of the Boy Pharaoh, Tutankhamen, who had been laid to rest 30 centuries before.

Although sequestered in a mountain to prevent the looting that afflicted the prominent pyramids, the tomb had been violated twice by robbers who left a cluttered confusion in their wake. Inexplicably, they also left behind riches of a like never before seen and scarcely imagined by the modern world. The glittering contents of the tomb provided vivid evidence of the ways in which early Egyptians used lavish amounts of gold and consummate artistry to guarantee immortality to their dead kings.

The labor of removing and cataloguing the more than 2,000 objects occupied Carter for 10 years. Intent upon recording every scrap of information the tomb offered, he took copious notes on the location and condition of each object before it was moved. He also commissioned an exhaustive photographic record, part of which is reproduced here, of the tomb and its splendid contents.

In the Valley of the Kings *(above)*, 33 royal tombs, including that of Tutankhamen, were found.

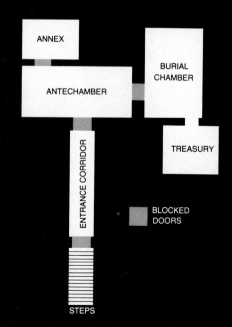

A diagram shows the layout of the burial suite.

ANNEX

BURIAL CHAMBER

ANTECHAMBER

ENTRANCE CORRIDOR

TREASURY

○ BLOCKED DOORS

STEPS

The tomb's antechamber contained a profusion of objects left in disarray by hasty thieves.

Carter and an assistant examine the innermost of Tut's three coffins. Solid gold and approximately six feet long, it weighed more than 240 pounds.

A gilded shrine housed Tut's internal organs.

A regal wooden bust of the King is escorted from the tomb in a procession led by Carter.

For all its grandeur, Tutankhamen's lavishly embellished golden throne portrays a moment of warmth and intimacy. Inlaid with colored glass and semiprecious stones, it shows a girlish Queen anointing Tutankhamen with perfume, possibly before his coronation. An awed Carter called the throne "the most beautiful thing that has yet been found in Egypt."

20,000 pounds of gold a year and even larger yields of silver. Caesar, as was the practice, kept a generous portion of the precious metals for himself and to pay his occupying army. After just one year he returned home with sufficient wealth to pay his debts and to buy himself election to the newly established triumvirate that would rule Rome.

The Romans' thirst for gold was unquenchable. In search of more, Caesar left Pompey to govern the city and invaded Gaul, while the third member of the triumvirate, Crassus, struck eastward with his army toward Persia. In Syria, Crassus was defeated and killed, but Caesar conquered the mines and storehouses of western Europe and the British Isles. He built thousands of miles of stone roads on which to transport his spoils home. Caesar's triumphant return to Rome in 46 B.C. was a testament to the political power of gold and silver. Prudently, he distributed a portion of his victor's booty among the soldiers and citizens of Rome, and they in turn acclaimed him dictator.

The profusion of precious metals in Roman daily life is difficult for modern men and women to comprehend. Great public buildings were trimmed and plated in gold. Golden statues proliferated. Both males and females adorned themselves unstintingly with gold and silver jewelry — in effect wearing their wealth for all to see. As the established currency of the realm, gold and silver passed from hand to hand, buying votes, ensuring the loyalty of armies, and purchasing the betrayal and murder of kings. Caesar, hoping to increase the Roman birth rate, offered rewards of gold to large families and forbade childless women to wear gold jewelry.

Taxes levied on Rome's far-flung provinces added to the enormous stores of precious metal taken by conquest. But the successors to Caesar proved better at squandering gold than at collecting it. Great sums poured back into the provinces to pay for the luxuries of the ruling class. Rome had no other commodity to exchange for the fine wines and dyed wool of Syria, Grecian marble or the silks and spices of the Far East.

The personal fortune of the Emperor Tiberius in the First Century A.D. is reported to have been the equivalent of more than $100 million. His heir, the demented Caligula, spent most of it during his brief four-year reign. After Rome mysteriously burned in 64 A.D., the notorious Nero set about rebuilding the city to his pleasure. A lavish personal residence of great size and luxurious content called the Golden House typified the extravagances of Nero's construction program; the walls were ablaze with gold and gems, and the bathrooms boasted gold fixtures. Nero's profligacy left the royal treasury perilously short of precious metals for use in currency. During his reign the silver denarius was debased by 20 per cent by adding copper, tin and lead. The percentage of silver used in the coin decreased steadily until, by the Third Century A.D., the denarius had become 98 per cent copper. The golden aureus also was degraded. In Caesar's time, the gold content had been guaranteed at $1/40$ of a pound, but by the reign of Domitian in the Third Century A.D., the gold content of the coin had been reduced by two thirds.

Tiberius had warned the Roman Senate against extravagances "that drain the Empire of its wealth and send, in exchange for baubles, our money to foreign nations and even to enemies of Rome." Eventually the distant mines that supplied Rome became exhausted. Over the centuries, the Romans had done almost nothing to modernize mining techniques to make the mines

This exquisite silver drinking horn, crafted in the shape of a deer's head, dates from the Fourth Century B.C. — the time of Philip II of Macedonia, father of Alexander the Great. Shaped by pounding a sheet of silver down into a hollow tree stump, the cup is typical of the metalwork of Thrace, part of Philip's empire.

more productive. The Empire's store of precious metals began to run out. Still the outward flow of currency continued; the self-indulgent life style it supported sapped the moral as well as the commercial strength of the Roman people. The decline was steady, and when the gold was gone, the Empire fell.

"Gold and civilization wax and wane together," wrote historian Will Durant. In the waning years of their Empire, the Romans dispensed their remaining store of gold and silver to hire mercenaries and to buy off the warrior tribes of so-called barbarians that threatened the northern borders of the Empire. Rather than circulate the precious metals in trade, the barbarians hoarded and buried them.

When the barbarians finally overran Roman Europe, operations in the few productive mines came to a halt. Within a generation after the Romans pulled back from Britain in the Fifth Century, for example, gold and silver coins ceased to circulate there. Starved for a medium of exchange, Europe sank into a commercial coma. By the Eighth Century, it had been reduced largely to an agricultural economy: Land was the sole source of wealth. Wages and rents were paid in personal services, and commodities were obtained through a system of barter.

The experience of Eastern peoples was far different. Gold and silver had always been abundant there. After the failing Roman Empire divided into eastern and western halves, the Fourth Century Emperor Constantine established a new capital on the Bosporus strait, east of ancient Troy. Constantinople, as this thriving city was called, became the seat of the Byzantine

Empire. As had their counterparts in Troy, the rulers of Byzantium controlled the trade routes to India and beyond. Their military and political power rested firmly on a tightly structured economic system sustained by a gold currency known as the solidus, or bezant. Circulated everywhere, the gold bezant, fixed in content, became the most prized coin in Christendom. On this golden base, Byzantium flourished for 800 years — far longer than the profligate empire of the Romans.

The reliable bezant paid the cost of Constantinople's defense against recurrent waves of marauding Persians, Arabs, Bulgars and Slavs. And, by the 11th Century, it provided the foundation for a remarkably complex and affluent society. Constantinople, wrote a Western contemporary, contained "two thirds of the world's wealth." The city's splendid thoroughfares were bordered by ornate palaces, magnificent monasteries and the villas of merchant princes. The wealthiest of its one million citizens lived a life of sophisticated luxury that no previous age had approached. Even the common folk, reported one visitor, "seem all to be the children of kings."

In medieval Europe, where even kings did not live particularly well, the example of Byzantium's flaunted wealth kindled a craving for a new source of gold and silver. This yearning led to a revival of the ancient practice of alchemy — the attempt to transform base metals into noble ones. Virtually every king employed at least one alchemist in the hope of finding the magic formula that would fill the royal treasury.

European alchemists experimented endlessly in their primitive laboratories, using heat, water and any number of catalysts in their elusive search. Many of them were mere charlatans, dedicated only to fooling their patrons. Others were fuzzy-minded magicians whose thinking ran heavily to superstition, exotic potions and the occult. Yet the best of the alchemists can be regarded as the world's first true practitioners of the scientific method. Archimedes, Pythagoras and the other early greats arrived at their principles and theorems largely through magnificent leaps of logic. But the alchemists were men of the laboratory and the experiment. Through patient years of trial and error they created the science of metallurgy, discovered new elements and invented new chemical compounds.

A Swiss alchemist, experimenting with quicksilver, found a remedy for syphilis. A German seeker of gold discovered instead how to produce crystalline antimony, which was used for making metallic alloys, such as type metal and pewter. Johann Böttger of 18th Century Dresden spent most of his career in alchemy as a prisoner of one selfish king or another. The masterwork that ultimately issued from the fires of Böttger's laboratory was not gold but a porcelain so fine that it has ever since been identified with his city.

For all their efforts, alchemists often were executed by their frustrated masters. Others were exiled in disgrace. Despite the threats that hung over them, not one succeeded in transmuting base metals into noble ones. As a sadder but wiser alchemist conceded on his deathbed, "To make gold, one must start with gold."

Conquest, not alchemy, ended Europe's shortage of gold and silver. The Crusades, mounted at first because of the religious zeal of feudal Europe to recapture the Christian Holy Land, became a vehicle for the plunder of the Byzantine Empire, still ruled from Constantinople. In 1204 a crusader force

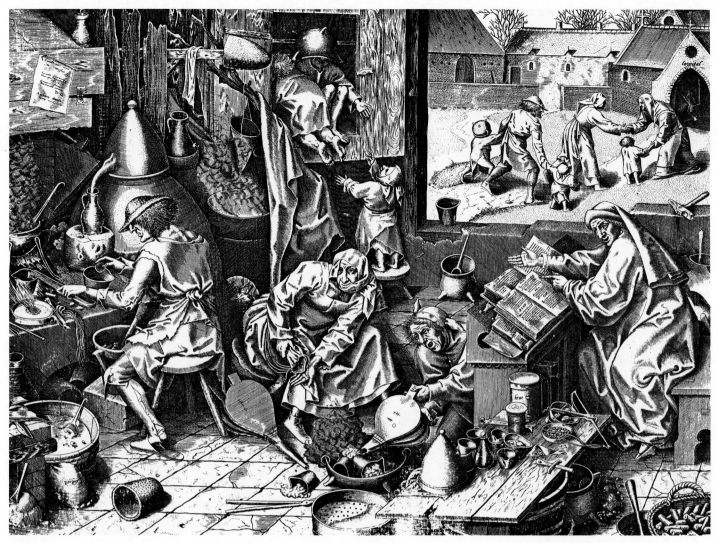

Alchemy's main product is depicted as misery rather than gold in this mid-16th Century etching by the Flemish artist Pieter Brueghel the Elder. A family works feverishly to manufacture gold under the direction of an alchemist seated at right; the window offers a view of their future as they are seen trudging to the poorhouse.

led by Venetians captured and sacked Constantinople. Their reward was the city's immense treasury and the Caucasus mining districts that supplied it.

The gold brought back from Constantinople made it possible to mint gold coins again in Europe after a lapse of centuries. The first of these was the florin. It was issued in the Italian city-state of Florence, which became a prosperous center for trade and artistic achievement. Just as the lack of gold had contributed to Europe's decline, so its return signaled the Renaissance.

The gold the crusaders brought back was merely a sample of the wealth to be found in the kingdoms to the east. The tales of the crusaders inspired Marco Polo of Venice to set out on his epic journey to Cathay and later moved the master navigators of the Mediterranean to seek a sea route to the Indies, by way of the Atlantic Ocean. One of the most determined of the mariners, Christopher Columbus, was particularly intrigued by Marco Polo's published description of a royal palace in Japan that had a roof of gold and chambers paved with gold plates.

Gold as much as anything else motivated Columbus in his scheme to sail around the world to the Indies. And a paucity of gold brought about his downfall. It was part of the Genoan's dream to secure enough gold to raise an army of 10,000 horsemen and 100,000 foot soldiers to conquer Jerusalem and the Holy Land in the name of Christianity.

This ambition fortified Columbus during the years of frustration when he could find no sponsor for his voyage of exploration, and it sustained him on the voyage westward across an empty sea, when his disheartened crew threatened mutiny. It was the promise of gold that had moved his low-born

sailors to sign on for the perilous expedition in the first place, just as it had persuaded the King and Queen of Spain to underwrite its cost.

The men who sailed with Columbus had visions of a land where gold was, as one of them wrote, "to be picked like ripe fruit from trees or fished from the river with nets." What they found instead in the Bahamas, and subsequently in Cuba and Haiti, were native Caribs wearing a few golden ornaments on their necks and wrists. The ornaments were made from nuggets of gold gathered in small amounts from stream beds. But there were enough of them to prompt Columbus to establish a colony on Haiti, having found there, he wrote, "such good will and such signs of gold!" The samples of gold Columbus brought back to Spain whetted the royal appetite. But when little more was forthcoming, the explorer fell into disfavor. A new colonial governor appointed by the King sent Columbus back in chains from his third voyage, in 1500, and though the explorer sailed once more to the islands, he lived out his last few years in poverty.

The immense New World treasures of gold and silver that made Spain the richest nation of the age were found by Columbus' hard-bitten successors — the conquistadors. Driven by a hunger for gold and glory, seeking as well to introduce Christianity to the Americas, these enigmatic freebooters left a trail of mission churches and plundered cities from Mexico to Peru. But few of them lived to enjoy the riches that they had wrested from the natives of the New World.

In 1517 Hernandez de Cordoba found shelter from a storm on the jungle shores of the Yucatan Peninsula and discovered the remains of the Maya nation, a civilization now believed to be as old as that of the Egyptians. The Maya were highly skilled in astronomy and mathematics — and in the forging of beautiful gold and silver images for their holy cities. Yet their elite metalworkers had never bothered to invent a metal digging tool for the farmers who fed them. Cordoba and his crew made off with the gold of the Maya temples, but not without a fight. Cordoba himself was wounded and soon died.

Three years later Hernando Cortez marched on the Valley of Mexico with 600 men, seeking more of the gold that Cordoba had seen on Yucatan. The Aztec, rulers of an inland empire rich in gold, mistook the bearded Castilian for a god whose return had been prophesied. Montezuma, the Aztec King, sent emissaries to Cortez bearing gifts of gold, including a disk, representing the sun, that was as big as a cart wheel. The messengers begged the Spaniards to turn back, but of course the sight of gold only spurred the invaders on.

The Aztec for two centuries had been levying gold tribute from subordinate Mexican peoples, who collected the gold by washing the gravel of streams in gourds. The temples and palaces of the Aztec capital, Tenochtitlan, were veritable storehouses of treasure. When Montezuma ventured out to meet the advancing Spaniards, he was borne on a golden litter.

By a combination of force and treachery, Cortez and his small army soon subjugated the mighty Aztec; the Spaniards then razed the Aztec capital, which Cortez himself had called "the most beautiful city in the world." The unfortunate Montezuma was killed; Cortez, in reward for the bounty he had shipped to Spain's King Charles V, was named governor of New Spain, but he was soon forced from office by rivals in the Spanish hierarchy. The conqueror of the Aztec died in 1547, broken and impoverished.

The gold-encrusted spires of the world's most famous Buddhist temple, the Shwe Dagon, dominate Burma's capital city, Rangoon. Golden offerings from the devout — monarchs and pilgrims alike — have been applied to the temple's roofs through the ages since it was originally constructed 2,500 years ago.

The scene suggests an epic movie about a bygone age. Swarms of men in search of gold are hacking at a mutilated mountain with picks and shovels and are struggling to carry away the debris in sacks. When backbreaking labor and blind luck conspire to reveal the glitter of gold in their pans or sluice boxes, the men cheer their lungs out in wild celebration. All this is taking place more than a century after the era of the independent prospector was supposed to have given way forever to the age of high technology and high finance.

The anachronistic story, set in the jungles of Brazil, began in January 1980 when a farm hand spotted a particle of gold in a stream. Word of the find soon leaked out, and one week later a thousand prospectors had staked out claims on Serra Pelada, or Bald Mountain, 270 miles south of the mouth of the Amazon River. They turned up nugget after huge nugget, and the gold fever spread; five weeks later the area population had topped 22,000. Armed men worked their claims by day and gathered at night along the "Avenue of Lies," the commercial district, to swap stories, gamble, drink whiskey and pay outrageous prices for necessities. The miners could afford it, for gold seemed everywhere. Some finds were mind-boggling; one prospector turned up a 15-pound nugget that was worth $108,000.

The long arm of the Brazilian government soon reached out to the boom town. Restriction after restriction was clamped on until by 1983, guns, gambling and alcohol had been outlawed. The government guaranteed mine workers a minimum wage, controlled prices of food and supplies, and established a monopoly on gold production — paying about 30 per cent less than the current world price. Serra Pelada's oldtime gold rush had been yanked at least partway back into the 20th Century.

Brazil's Serra Pelada teems with gold miners hacking away at the sandy rock of their claims and carrying the mountain away by the sackful for sluicing. In less than four years, these archaic methods yielded 30 tons of gold.

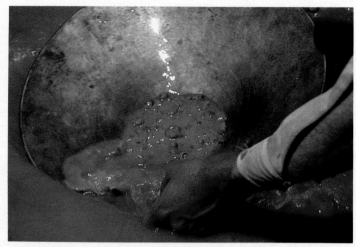

Gold is sifted from the rich workings of Brazil's Serra Pelada mine with implements that were last in widespread use during the legendary gold rushes of the 19th Century. Although the sluice box *(left)* can process dirt more quickly than the simple pan *(above)*, both methods are laborious and time consuming.

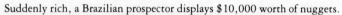

Suddenly rich, a Brazilian prospector displays $10,000 worth of nuggets.

The source of the enormous metallic wealth of the Americas was still to be found, and the search would give rise to the persistent legend of El Dorado — the Land of Gold. It began as the tale of an Indian chieftain who, once a year, anointed himself with a sticky substance and rolled in gold dust before bathing in a sacred mountain lake. The story, and the quantity of gold involved, expanded with each retelling. Indeed, as gold fever quickened, El Dorado became as much a state of mind as an actual place to be discovered.

The Indians were all too aware of the European thirst for riches. When pressed by Spanish adventurers for information about the sources of gold, they always pointed southward. The prospect of finding a kingdom of gold beyond the southern horizon drove Vasco Nuñez de Balboa and a small band of men across the untamed and disease-ridden Isthmus of Panama in 1513. Balboa discovered the Pacific Ocean, but before he could organize a new expedition to explore South America's Pacific shore, he was arrested by a rival for power, tried for treason and beheaded. But one of Balboa's band, an illiterate swashbuckler named Francisco Pizarro, went on to discover the golden kingdom of the Inca, in a high valley of the Andes Mountains in what is now Peru.

The Inca had been working with gold for 2,000 years. Remarkably like the ancient peoples of the eastern Mediterranean, they worshipped the sun as the father of life. Gold was the Inca's holy metal; they believed it emanated from the sun. They gathered it not only from rivers but from mountain mines as well. It had no value to them as currency — they used a system of barter instead. But for centuries Inca goldsmiths had been fashioning the holy metal into idols and helmets, pots and cups, and plates large and small, all made of pure gold. Some of the items were truly dazzling, for the artisans had learned to weld gold wire to larger bits of gold to create wonderfully intricate patterns.

The ultimate tribute to the Inca god was a massive sun temple. Made of carefully cut and fitted stone, with steps that seemed to reach the sky, the temple was revered as the residence of the sun god, Inti, whose earthly likeness in the temple had a human face, fashioned from a mask of beaten gold. Around him, cornstalks, animals and lumps of earth carved from pure gold were meant to assure the Inca farmers that Inti would warm the fertile Andean soil and provide a rich harvest.

In 1532 Pizarro rode inland from the Peruvian coast at the head of 368 men armed with swords, crossbows and a few muskets. Atahualpa, the Inca King who — like the pharaohs of Egypt — was worshipped as a divine offspring of the sun, came out to meet him with an escort of 5,000 warriors. The King was borne majestically forward seated on a golden throne; a string of emeralds around his neck was offset by a breastplate of glittering gold. His attendants were unarmed. After a false show of friendship, Pizarro suddenly signaled his men to attack. Within 30 minutes, and without losing a man, the Spaniards slaughtered 2,000 of the Inca and took the rest prisoner, including Atahualpa.

Held captive in a room 17 feet long and 12 feet wide, the Inca King offered the greatest ransom in history for his freedom. He would fill the room with precious metals as high as Pizarro could reach. For two months the subjects of the King brought forth gold and silver artifacts of every shape and size. The Spaniards' only interest in the metal was economic. No

Getting More Glitter from Less Gold

As long as 1,500 years ago, Indian metalsmiths in Central and South America had developed remarkably sophisticated techniques for making objects consisting mostly of copper appear to be pure gold. A skill the alchemists would have envied, the ancient process of depletion gilding diagramed at right involved removing from the object's surface, with applications of heat and chemicals, everything that was not gold. When he saw an example of their work in 1520, the German artist Albrecht Dürer marveled "at the subtle ingenuity of the men in these distant lands."

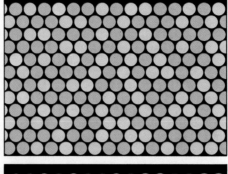

A diagram represents the atoms of an alloy, used by Chimu metalworkers in Peru, of copper *(tan dots)*, silver *(light gray dots)* and gold *(yellow dots)*. The alloy melted more easily than any of its constituents and was thus easier to work.

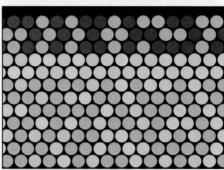

When the object has been shaped, the metal is heated until the copper on its surface combines with oxygen in the air. The oxidized copper *(dark gray dots)* forms a black, scaly coating.

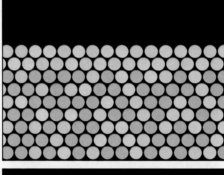

Next, the metal is bathed in a mildly acid solution of urine or plant juices. The outer coating of oxidized copper dissolves, leaving a surface layer of silver and gold.

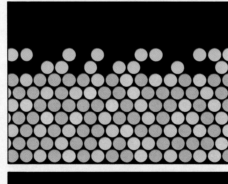

A highly acid paste is used to remove the silver particles from the surface of the object. The metal that remains on the surface is pure gold, but it is rough and porous.

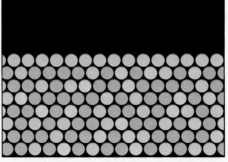

The final step is to heat the gold again and polish the surface. The topmost particles are thus smoothed into a glittering golden sheath for the sturdy alloy beneath.

For all its golden gleam, this figurine is, in fact, made of tumbaga, a gold and copper alloy. Quimbaya Indians of Colombia used depletion gilding to create its gold surface.

matter how beautiful the objects, they were melted down into ingots that could more easily be transported and divided among the conquerors.

In all, the ransom amounted to 13,000 pounds of gold and 26,000 pounds of silver, or roughly 100 pounds per man. But it was not enough to save Atahualpa. To deprive the Inca of their leader, Pizarro ordered him strangled in full view of his people. A similarly violent fate awaited most of the conquistadors. Soon they were fighting bitterly among themselves over gold and power. To consolidate his rule, Pizarro founded a City of Kings (which became Lima) and erected an impressive palace there for himself. But one of his original confederates, another illiterate swashbuckler, named Diego Almagro, rose against him. Defeated in battle, Almagro was beheaded. But his son assumed leadership of the insurgent faction and carried on the fight. In 1541, only a decade after he landed in Peru, Pizarro was assassinated by his fellow Spaniards in his own palace.

The hoard of gold and silver seized from the Aztec, Maya and Inca peoples was but a suggestion of the immense metallic treasure that was subsequently carried eastward across the Atlantic to Spain. From Mexico south, the Spanish established a network of mines that were worked by the Indians they had subjugated. The output of the mines was dispatched in great convoys that sometimes numbered more than 100 ships. Ocean storms and enemies — many of them privateers commissioned by rival governments in Europe — claimed numerous treasure ships. But many galleons made the crossing safely. In the 1550s alone, Spain was enriched by 100,000 pounds of gold from its possessions in the New World.

Spain became the first world power to develop the economic and political doctrine known as mercantilism, which became the bridge between the feudalism of the past and the capitalism of the future. A basic principle of mercantilism was to accumulate bullion in the national treasury and with it to finance the armies and navies that could intimidate rival powers and control foreign trade. Controlling trade and crippling the economic capacity of other nations became an even more critical goal after rival nations — notably Britain, France and Holland — followed Spain's lead in the quest for overseas empires. Preparation for war became a permanent part of every government's mercantilist policy.

Britain's defeat of the Spanish Armada in 1588 was followed closely by the establishment of English colonies in North America. Britain's rulers hoped to find gold and silver in the same abundance their Spanish enemy had discovered farther south. In this they were disappointed, as were the French in Canada. The North American colonies produced lumber, furs, tobacco and other commodities for the home country — as well as a captive market under the rules of mercantilism — but almost no gold or silver.

Britain emerged from the Seven Years' War in 1763 as the supreme power of the mercantile era. But its treasury had been drained, and the cost of maintaining troops and ships in the Western Hemisphere promised to remain high. The economically oppressive steps taken in London during the next decade to remedy this situation would eventually lead to colonial rebellion.

At the behest of King George III, Parliament passed a series of laws restricting the freedom of the American colonies to trade with any nation except Britain, and forcing colonial merchants to trade only on terms lopsidedly favorable to the mother country. One of the laws, the Currency Act of 1764, prohibited the colonies from issuing paper money. The Americans had no native resources of gold and silver, and the outlawing of paper currency left them hard pressed to do business at home, much less to discharge the debts they already owed to British lenders.

The desperate shortage of hard coinage not only contributed directly to the onset of revolution, but seemed to foredoom the rebels' cause when war began. The Continental Congress issued paper money, unsupported by bullion, that set off runaway inflation. By 1780, a barrel of flour to feed George Washington's army cost a staggering 1,575 continental dollars. Many merchants refused any payment other than gold or silver for their wares; with so little sound currency available, commerce in the colonies lapsed into a largely barter system.

Gold from England's European rivals eventually saved the American independence movement. France loaned the struggling colonies $6.4 million in gold and silver; Holland and Spain lent lesser amounts. Ironically, in helping to save the American cause, France depleted its treasury to a dangerous level—one factor among many that led to the overthrow of the Bourbon monarchy in 1789.

The newborn United States, meanwhile, faced an uncertain future because of its persisting shortage of precious metals. But ambitious American merchants and sea captains, freed from British restraints, set about solving the problem. While Britain and France were preoccupied with the Napoleonic Wars, American trading ships sailed the world, bringing home handsome profits in gold and silver. By 1820 the U.S. government had paid off most of its foreign debts in hard currency.

Until almost the middle of the 19th Century, little gold had been mined in the United States: The only significant strike had occurred in North Carolina in 1799, but it failed to produce appreciable amounts of gold. Then, early in 1848, the United States concluded a successful war with Mexico by annexing a vast tract of Western land, from Texas to California, in return for the payment to Mexico of $15 million. Unbeknownst to the negotiators of either nation when the treaty was signed, gold had been found nine days earlier at a sawmill on the American River near Sacramento, California. The discovery at Sutter's Mill and subsequent finds nearby set off an epidemic of gold and silver fever that shaped the development of the United States for the next half century and helped the young nation to withstand its greatest test.

In the first years of the great California gold rush, more than 100,000 men and women from all over the world made their way by land and sea to what had been a sleepy Spanish province. By the end of 1852 most of the easily worked fluvial deposits had been exhausted. But the exploitation of a vast mother lode of surface and underground gold had only begun. In the decade after the Sutter's Mill discovery, California produced more than half a billion dollars' worth of gold—almost 35 times the amount the United States had paid to Mexico for the Western lands.

More important, the rush to California accelerated the migration that made the United States a truly continental nation. And the glittering promise of the West attracted foreign investment in the still-underfinanced country. British investors, for example, helped to capitalize a boom in American railroad building. In 1850, foreign investment in the United

States totaled more than $500 million. By 1860 that amount had doubled.

The sudden attractiveness of the Western lands also fueled a bitter sectional controversy that had threatened the American republic since its founding. The question of whether to allow the extension of slavery into the Western states and territories had simmered through decades of compromise. Thirteen years after the discovery of gold in California, the dispute exploded into civil war.

In this war, like so many others before it, the availability of gold and silver played a crucial part. The Southern Confederacy, throughout its brief existence, suffered a chronic lack of precious metal to back up its paper money. This shortage severely limited the Southern government's ability to purchase the arms and even the food it desperately needed to survive. Confederate money became a synonym for something of little value. The North, by contrast, benefited from a steady flow of bullion from the new mining regions in the West that made its paper money "as good as gold." Between 1861 and 1864, a total of $186 million in gold and silver was shipped to the Union states, enabling them to meet their obligations promptly and with confidence. "One can only guess," mused mining engineer and author Thomas A. Rickard, "what might have happened if this precious bullion had gone to the South instead."

The California bonanza and subsequent strikes in half a dozen other Western states transformed the United States into a leading producer of new gold and silver, and vaulted the young nation to a place among the world's economic powers. It was an era of industrialization and expanding trade — both of which were linked to the availability of gold.

Fortunately, more gold was available than ever before. The California gold rush was followed by a rush to Australia in 1851, the discovery of fabulous gold fields in South Africa beginning in 1886, and rich strikes in Alaska and the Canadian Klondike in the 1890s. Within a quarter of a century, the world's supply of gold doubled.

A prime beneficiary of this was the British Empire. Determined to stabilize its economy after the costly Napoleonic Wars, Britain became the first nation to commit itself to a monetary system known as the gold standard. It guaranteed that Britain's paper money, the pound sterling, was redeemable in gold on demand. Gold was free to circulate among individuals and nations as the universal currency, and the amount of paper money in circulation was controlled by the amount of gold available to back it up.

The gold standard had the dual effect of instilling confidence in paper money and regulating the pace of growth. Gradually, other governments followed the lead of the wealthy, stable British, and by the beginning of the 20th Century, nearly 30 nations — including even the aggressively expanding United States — had subscribed to the gold standard. Mexico and a few other Latin American countries maintained a silver or bimetallic standard, but for the most part it was gold that both fueled and controlled the world's economies.

But the new century brought profound and increasingly rapid change. The first of its several wars was so cataclysmic and costly that it would sweep away the old order of things — including the gold standard. And the rapid-fire advances in science and technology bursting on the modern world transformed not only the uses, but human understanding of the sources of all the noble metals. Ω

THE MINTED WEALTH OF NATIONS

Passed from hand to hand and from century to century, ancient coins of gold and silver still bear bright witness to the early coinmakers' art — and to the abiding human need for currency.

The first true coins — their weight and value guaranteed by a government — originated around 550 B.C. in the kingdom of Lydia, which is now part of Turkey. Soon each of the Greek city-states was issuing its own coins, and coinmaking became a major art form.

The glittering masterpieces shown here and on the following pages, enlarged for detail and accompanied by actual-size photographs, began as measured lumps of rough metal. Heated to a red glow, they were placed between two deeply engraved dies. Under the blows of a heavy hammer, the dies imparted their images in relief to each of the coins' sides.

That many of these coins remain in near-perfect condition is due in part to the durability of gold and silver and in part to a comparably durable human trait: Some of the best specimens were buried by people seeking to preserve their wealth and unearthed centuries later as part of long-lost treasures.

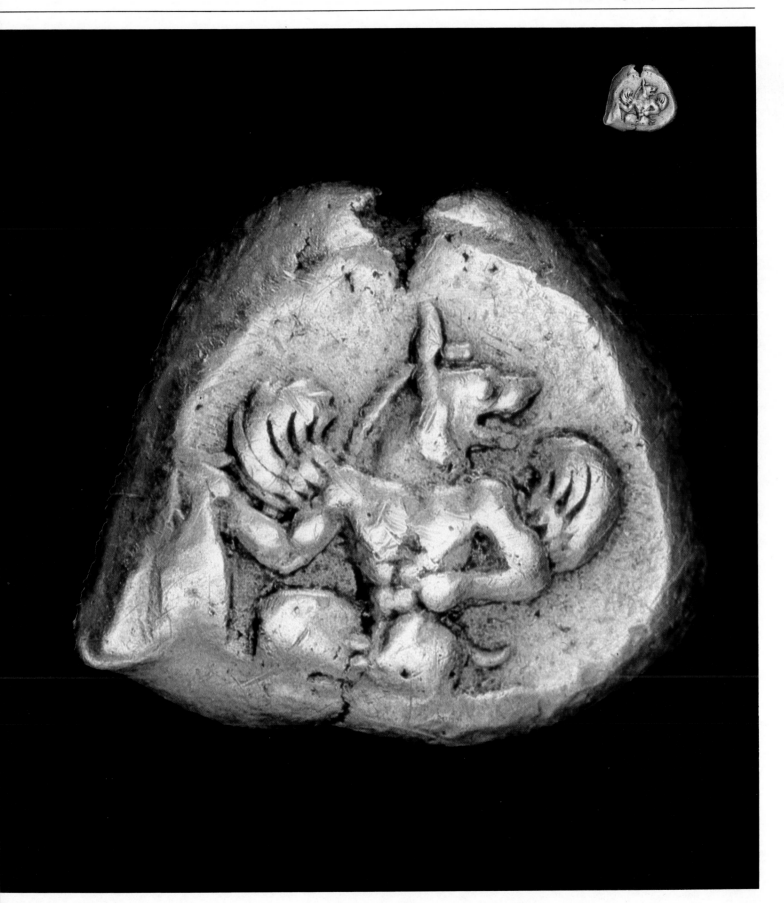

A great-eyed owl, the symbol of Athens,
stares from a silver tetradrachma struck around
440 B.C. The coin weighs four drachmas, a
standard unit of weight. This design remained
virtually unchanged for 400 years.

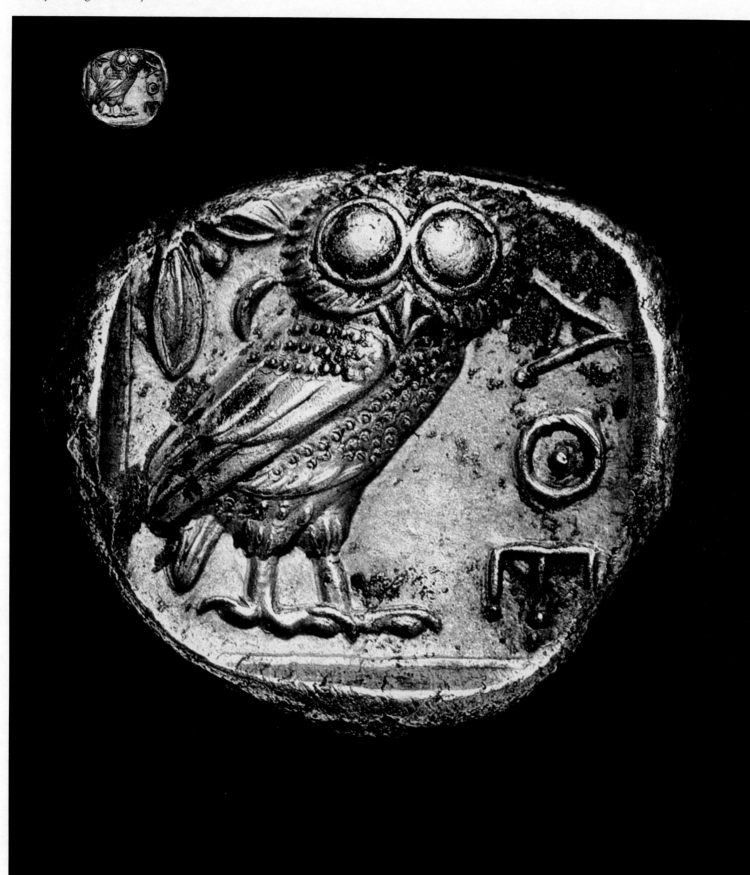

Rightfully proud of his work, an engraver named Kimon signed the die for a silver decadrachma around 400 B.C. The coin, from the Greek city-state Syracuse, bears his signature on a dolphin and his initial on the nymph's headband.

Coins struck in the waning years of the
Roman Republic often bore the likenesses of
generals and were used to pay their troops.
This gold aureus depicts the rugged features of
Brutus — one of Julius Caesar's assassins.

Barred from issuing coins while under Roman domination, Jewish citizens of Palestine did so during periods of independence by restriking Roman coins. The example below was fashioned around 134 A.D. from a silver denarius.

The gold solidus or bezant — this one struck around 614 A.D. — became the staple currency of the Eastern Roman Empire, which flourished after the fall of its Western counterpart in the Fifth Century A.D.

This silver English penny dates from the late 10th Century. For 500 years during the Middle Ages, the penny was England's sole coinage. When smaller denominations were needed, they were created by cutting the coins in pieces.

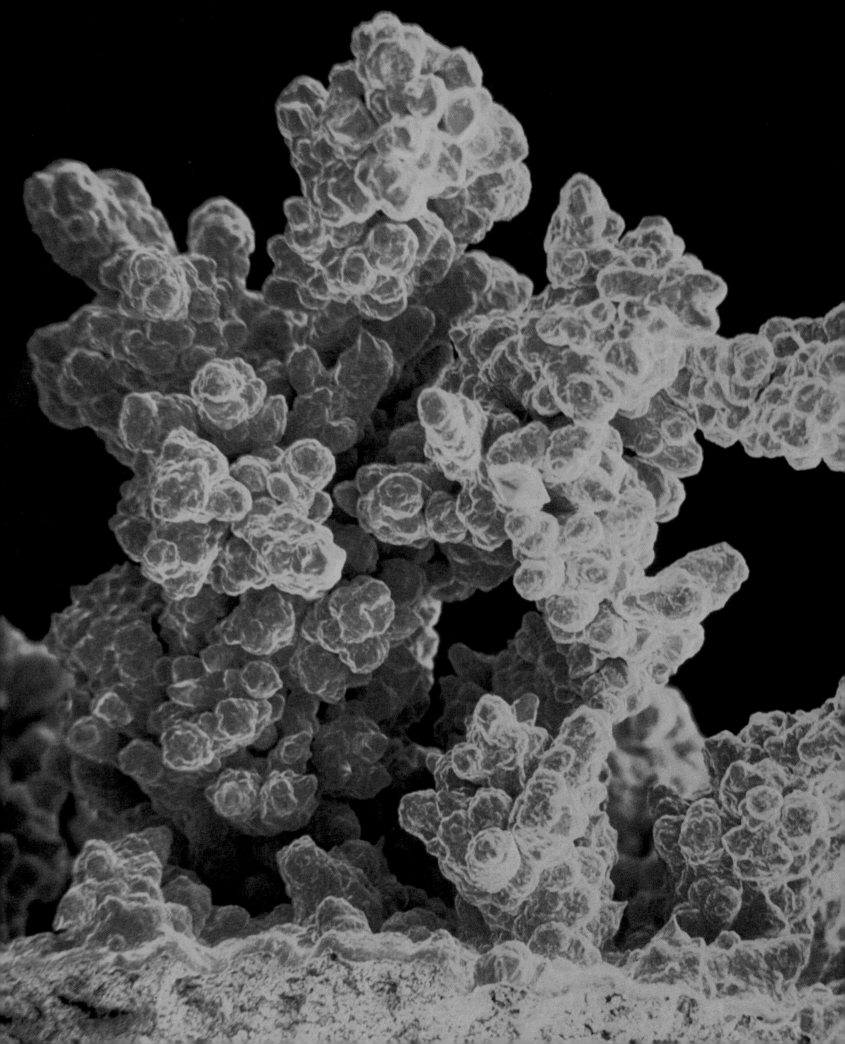

Chapter 2

THE GENESIS OF MINERAL TREASURES

Humanity's never-ending search for precious metals has been a galvanizing force in the developing science of geology. Yet for all the fierce economic motivation underlying it, the study of the earth and its secret workings has been a late-blooming science. Seldom has a subject been so shrouded in myth, magic and misconception. And seldom have the errors committed by ancient scholars persisted so powerfully into modern times. The first published study on mining and mineralogy based substantially on scientific observation rather than mere speculation did not appear until the 16th Century. Indeed, most of the success in unraveling the earth's geologic mysteries has been achieved only in the past 100 years. The work continues, spurred by a voracious demand not only for gold, silver and platinum but for all the useful minerals on which industrial society is based.

A Biblical view, based on the Book of Genesis, held that metallic deposits originated spontaneously at the moment of creation. The Greek philosophers believed that metals were living things that grew and propagated themselves in the rocks of the earth's core. They reasoned that if the other two kingdoms of nature — animal and vegetable — reproduced themselves by means of seeds given off by their own bodies, then minerals must do the same. By this logic, it followed that if a worn-out ore field were given a sufficient rest, it would replenish itself and become worth mining again.

A corollary to the ancients' belief in "living" metals was the notion that the content of the earth's crust was gradually and constantly changing, from baser metals into more valued ones. Aristotle, in the Fourth Century B.C., wrote that "nature always strives after the better." By that dictum, lead ore containing some silver was thought to be in the process of transmutation to pure silver; samples of silver that contained traces of gold, as silver often did, were maturing into gold. So it was argued that if a silver mine were shut down for a generation or two, it could be reopened as a gold mine.

The astrologers, for their part, contended that metals were directly influenced by heavenly bodies. The characteristics of each of the seven known metals were ascribed to one of the seven major celestial bodies that were believed to revolve around the earth. They in turn were closely identified with individual gods. The lively properties of quicksilver, for example, were associated with Mercury, the nimble messenger of the gods and the swiftest of the planets. Iron, the essential stuff of warfare, was believed to be influenced by Mars, the red planet named for the god of war. Lead, which was dense and dull, was identified with Saturn, which was the farthest known planet from Earth and appeared to move the most sluggishly. Gold

Crystals of almost perfect gold, its impurities refined away by electrolysis and its color enhanced by computer, are shown magnified 300 times. Called "five-niner" gold because it is 99.999 per cent pure, such gold is too soft for most practical uses and will have to be alloyed with other metals for hardness.

and silver, of course, were inextricably tied to the sun and the moon.

Aristotle attributed the geologic make-up of the earth to the power of the sun. He declared that the rays of the sun caused exhalations that penetrated the earth's crust and made possible new combinations of elements, created metals and minerals, and formed various kinds of stones.

Remarkably, Aristotle's doctrine remained unchallenged for the better part of 2,000 years. It was supported with little change by such intellectual giants of the Middle Ages as Albertus Magnus, Roger Bacon and Saint Thomas Aquinas. As late as the 16th Century, common wisdom held that metals were created by rays emanating from celestial bodies. "Every metallic ore receives a special influence from its own particular planet," wrote the respected German metallurgist Ulrich Rühlein von Calbe in 1527. "Thus gold is of the Sun or its influence, silver of the Moon, tin of Jupiter."

The man who firmly and finally set aside the theories of the scholastics was a German physician named Georg Bauer who wrote in Latin, the universal language of his day, under the name Georgius Agricola. Born in Saxony in 1494, Agricola was educated in Italy during the full flowering of the Renaissance and settled in the booming German mining district called the Erzgebirge, or Ore Mountains. A true Renaissance man, Agricola was a political leader and diplomat as well as a physician and scientist. He wrote widely on subjects as disparate as the bubonic plague and the weights and measures of ancient Greece and Rome. But his major contribution was a series of books on mining and metallurgy, climaxing in a masterwork entitled *De Re Metallica* that was published a year after his death in 1555 and established the Saxon doctor as the father of modern geology.

What set Agricola apart from the thinkers before him was that he was an energetic investigator and observer, not a mere logician. "Those things which we see with our eyes and understand by means of our senses are more clearly to be demonstrated than if learned by means of reasoning," he wrote. Following his own advice, he spent much of his time prowling the silver, gold and lead mines of the Erzgebirge and questioning the workers.

Profusely illustrated with woodcuts, *De Re Metallica* was the most comprehensive treatise ever assembled on mining history and techniques. Though not wholly accurate, it remained useful well into the 20th Century. Agricola outlined the many ways — from avalanches to a farmer's plow to the taste of water — in which metallic-ore deposits might be discovered. He ridiculed the superstition that gold or silver could be located with a forked stick called a divining rod. And he rejected the theories of both Biblical scholars and astrologers. Agricola also found unacceptable Aristotle's view that the sun's exhalations were the origin of metal and valuable stones.

Instead, Agricola looked to natural sources, particularly to the solution of minerals in liquids and their precipitation by means of gravity, heat, cold and evaporation. He explained for the first time the role of erosion in the shaping of mountains and the formation of ore deposits, which he said evolved later in time than the rocks that contained them.

Agricola's most original contribution was his suggestion that erosion caused by subterranean waters created breaks in the earth's crust and that ores were deposited from solutions that circulated in these openings. Agricola's proposition was fundamentally correct, as far as it went, although his understanding of what the underground solutions — he called them juices — were composed of was little better than that of the ancient Greeks.

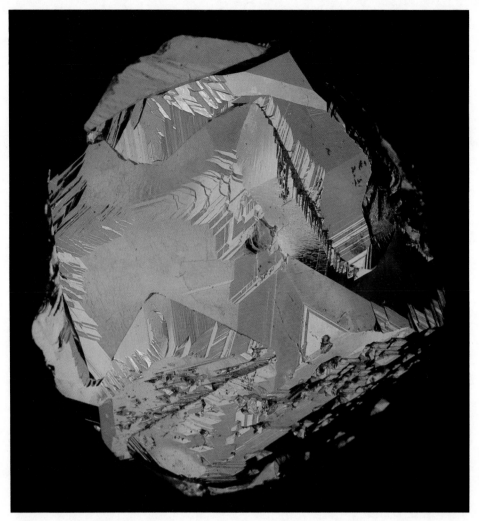

Crystals of pyrite, an iron sulfide, display the metallic luster and color that account for the metal's nickname — fool's gold. Despite the disparaging name, pyrite is not always worthless; its presence is a reliable clue to the existence of a vein of gold nearby, and it sometimes contains enough gold compounds along with the iron to make it a worthwhile ore.

Nor was the Saxon doctor immune to all the fantasies commonly accepted in 16th Century Germany. One chimera had it that bands of frightful demons lived in the mines and preyed upon the miners from time to time. Agricola did not deny the demons' existence. However, he explained that the little fellows were more mischievous than dangerous; they stood only two feet tall, and unless one made fun of their modest stature, they would do nothing more than throw pebbles at the humans who invaded their underground world.

Despite such lapses into fancy, Agricola's masterwork proved to be a bridge between ancient shibboleths and modern scientific investigation. Generations of researchers, building upon his work, have constructed a detailed picture of the history and nature of metals in general, and the noble metals in particular. Yet even today geology is anything but a static body of knowledge. Scientific concepts of the subterranean forces that shape the world are subject to constant challenge and change.

Since Agricola's time, physicists have determined that all solids, liquids and gases are composed of minute particles of matter called atoms. They have identified more than 100 fundamental elements, each of which is made up of a different kind of atom. The elements rarely occur in pure form, but are usually found in natural compounds of two or more, called minerals. Almost 3,000 different minerals can be distinguished from one another by individual characteristics that result directly from the kinds of atoms they contain and the arrangement of the atoms inside them. In fact, every property of matter is a product of the patterns formed by its atoms and the strength of the forces that bind the atoms together.

The atoms in minerals tend to arrange themselves in distinctive geometric patterns called lattices that repeat regularly to form crystals. The effect can be compared to a wallpaper pattern, except that lattice patterns repeat in three dimensions rather than just two. The several different forces that cause atoms to combine and stay together are called atomic bonds, and they are responsible for many of the qualities that make each mineral unique.

The noble metals have a number of qualities that set them apart. One is rareness. Gold constitutes only 3.5 parts per billion of the earth's crust, platinum about 45 parts per billion and silver 73 parts. All three of these metals are strong and relatively heavy as a result of their close-packed atomic structure. Gold's density, 11 ounces per cubic inch, is exceeded only by certain members of the platinum group. Silver ranks directly behind gold at 6.25 ounces per cubic inch. The malleability of gold and silver, a joy to metalsmiths through the ages, also helps them to survive the destructive atmospheric forces to which they are exposed in nature. Bonds between metal atoms are unique in that electrons in the individual atoms merge until the entire mass is a group of atomic nuclei spaced at regular intervals in a sea of swimming electrons. Malleability is a result of this free electron movement, which allows entire blocks of atoms to slip easily in opposite directions when a strong force is applied. The atoms tend to rebond once the force is removed, and the metallic mass remains in one piece.

Of the three noble metals, gold is the softest, and for this reason it is the most often alloyed with other metals to harden it for commercial uses. Platinum's extreme natural hardness makes it an ideal jewelry setting for the hardest of all minerals, diamond. Platinum's steely gray appearance, however, falls somewhat short of the dazzling brilliance of either gold or silver — which is one reason why it was not prized as a precious element in its own right until the 19th Century. The high luster of all three metals is caused by the mobile electrons within the atoms that interfere with light rays and reflect them with special brilliance (*pages 55-57*).

Gold and platinum are among the relatively few metals that are commonly found in their native, or pure, state, uncombined with any other elements. Silver, which of all metals boasts the highest capacity to conduct heat and electricity, is most often found in combination with other elements such as gold, sulfur and chlorine. Silver in fact is an element in about 50 different mineral species.

But the most distinctive quality of the noble metals is the relative chemical inertness that makes them almost impervious to the corrosive forces of moisture, common acids or alkalies. If there is a vulnerable sister among the three it is silver, which is stable in pure air and water but tarnishes in the presence of sulfur, hydrogen sulfide or ozone. Even this tarnish — as anyone who has ever polished silverware knows — can be removed by applying an alkaline solution that restores the silver to its natural moonbeam color.

Since ancient times, the purity of gold has been measured in terms of karats, an arbitrary scale on which pure gold has a rating of 24 karats. Thus 12-karat gold is 50 per cent pure. The term should not be confused with the carat used as a measure of weight for diamonds and other gemstones. (Both words derive from the Italian *carato* and the Arab *quirat,* which mean "fruit of the carob tree." The seeds of the fruit, remarkably uniform in weight at

The Atomic Secrets of Precious Metals

Many of the practical and esthetic qualities that set the noble metals apart from other elements can be traced to the independent performance of a single electron, the tiniest of all elemental particles. The atoms that constitute every element contain a specific number of negatively charged electrons whirling in roughly concentric orbits called shells around a cluster of protons and neutrons called a nucleus. (An atom of gold, depicted below, has 79 such electrons in orbit; platinum has 78 and silver 47.)

Those electrons in the inner orbits are relatively stable and are considered "closed." Those in the outermost, or valence, shells are less attracted to their nucleus and are considered "free" to roam and to bond with other atoms.

The freest of all electrons are the solitary ones that orbit farthest from the nucleus in each atom of gold, platinum or silver. Bound to no single atom, and moving at extremely high speed, these free electrons are shared by all the millions of atoms that make up a piece of the metal. The wandering particles in fact form a kind of electron gas whose mobility and other characteristics (*following pages*) endow the precious metals with their malleability, their capacity to conduct electricity and heat, and even their renowned luster.

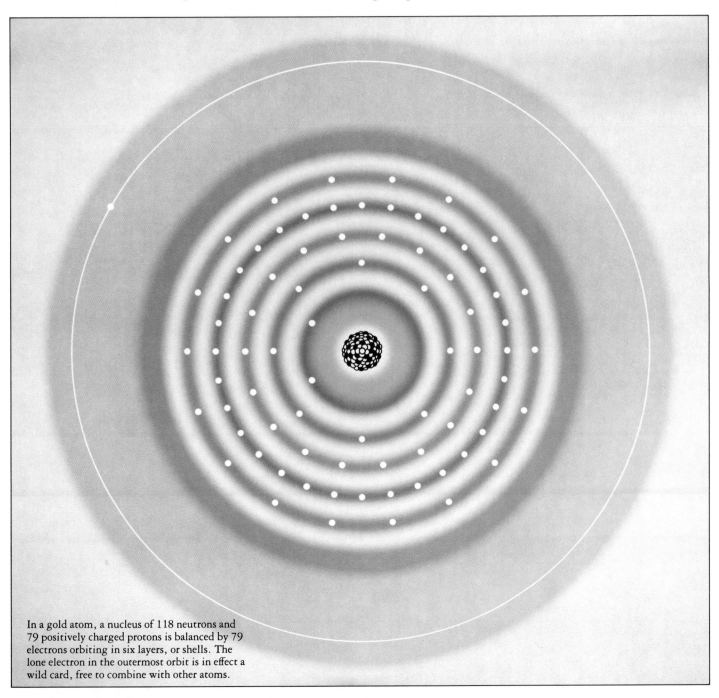

In a gold atom, a nucleus of 118 neutrons and 79 positively charged protons is balanced by 79 electrons orbiting in six layers, or shells. The lone electron in the outermost orbit is in effect a wild card, free to combine with other atoms.

A BONDING OF ELECTRONS

The atoms in metals are almost always packed closely together and are bonded by the sharing of electrons between two atoms. This adherence is called a chemical bond. But in each atom of gold *(right)* and other noble metals there is a single outermost electron so loosely attracted to its nucleus that it is free to travel *(arrows)* and form temporary liaisons — metallic bonds — with all the atoms in the metal.

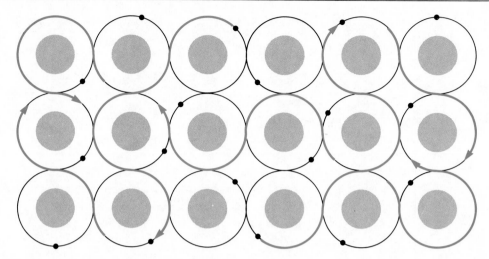

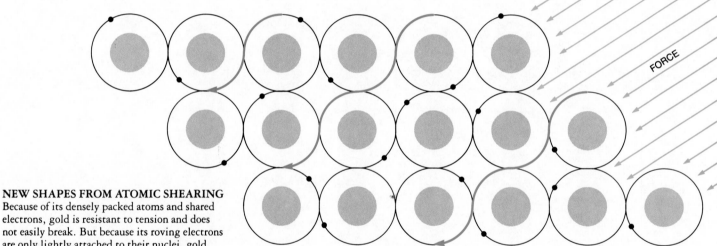

FORCE

NEW SHAPES FROM ATOMIC SHEARING

Because of its densely packed atoms and shared electrons, gold is resistant to tension and does not easily break. But because its roving electrons are only lightly attached to their nuclei, gold atoms tend to shear away from one another when force is applied. This accounts for the ease with which the noble metals can be hammered into new forms or extruded into thin wire. Once the force is removed, the metallic bond is renewed and the metal retains its new shape.

A CONDUIT FOR ELECTRICITY

Electrical conductivity in metals is defined as the movement of free electrons in a preferred direction. When placed in an electrical field, negatively charged electrons are repelled by the negatively charged side of the field and tend to flow toward the positive side. Because of their mobility, the free electrons whip through gold, silver and platinum at great speed, facilitating the flow of electrical current.

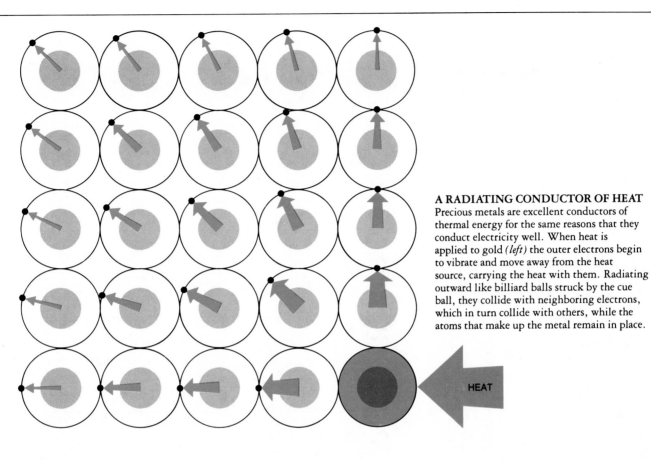

A RADIATING CONDUCTOR OF HEAT
Precious metals are excellent conductors of thermal energy for the same reasons that they conduct electricity well. When heat is applied to gold *(left)* the outer electrons begin to vibrate and move away from the heat source, carrying the heat with them. Radiating outward like billiard balls struck by the cue ball, they collide with neighboring electrons, which in turn collide with others, while the atoms that make up the metal remain in place.

HEAT

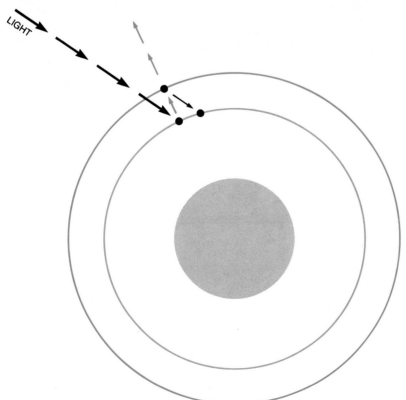

LIGHT

THE SOURCE OF LUSTER
The eye-pleasing radiance that is associated with the noble metals is a result of teamwork between light and the wild-card electron. When struck by light *(inbound arrows),* the electron absorbs energy that enables it to jump into a higher, faster orbit around its atomic nucleus. But the new orbit cannot be permanently sustained — the attraction of the nucleus is stronger than that of the light. As the electron falls back into its original orbit, its absorbed energy, no longer needed, is radiated outward in the form of light and is visible as luster.

about one fifth of a gram, were used to balance the scales when weighing gold and gemstones in ancient bazaars.) Gold is also rated on a scale of fineness, which ascribes the value "1,000 fine" to pure, or 24-karat, gold.

The color of pure gold is a rich yellow. The more silver it contains, the lighter its color, and even a modest amount of copper gives gold a reddish cast. Pyrite, a common mineral of little value, has a similar color. Over the years, the similarity has raised the hopes of many an inexperienced prospector, until he realized that he had found nothing more than fool's gold.

Those compounds in the earth's crust from which valuable metals can be extracted are known as ore minerals. The definition of ore is strictly commercial: It is any mixture of ore minerals and worthless material, called gangue, from which metal can be recovered at a profit. Thus the market value of the metal in the mixture, as well as the cost of mining and transporting it, determines whether or not it is ore. For example, a mineral deposit that contains 10 pounds per ton of copper, an abundant, low-cost metal, would barely qualify as ore. But if the same ton contains as little as one fourth of an ounce of gold it is considered a rich find.

A single ore deposit may yield several different metals. Because gold and platinum commonly occur in their native state, uncombined with any other elements, their separation from the gangue and from lesser minerals is a relatively simple process in which the mechanical forces of nature often have done much of man's work for him *(pages 74-83)*.

All mineral deposits, including those rich in gold, silver and platinum, lie on or within the rocks that make up most of the earth's crust. The elements were formed at the same time as the earth itself, which more than 4.5 billion years ago gradually cooled from a gaseous mass into an externally solid sphere. But the concentration of the elements into ore deposits, and their transport to or near the earth's surface, took place more recently, just as Agricola contended. Some deposits have been in place for hundreds of millions of years; others are still forming.

The rocks that play host to mineral deposits are divided into three general groups: igneous, sedimentary and metamorphic. Igneous rocks are the product of molten material called magma (meaning "paste" in ancient Greek), a fiery chemical solution that originates deep within the earth. The magma consists of molten minerals, water vapor, gases and small amounts of valuable metals. Driven toward the surface by mighty subterranean pressures, the magma makes its way upward by melting the rocks above it or, near the surface, by filling existing cracks and crevices. The magma slowly cools, loses its gases and vapors, and crystallizes into solid rock. On occasion the magma erupts through to the surface as a volcano and is called lava. But most magma cools and hardens well below the earth's surface.

Sedimentary rock is formed when rock fragments are transported by water, wind or glacier from their original source on or near the earth's surface and redeposited at a new site. They accumulate in gravel-like conglomerates, fine-grained sands or muds, and eventually they are consolidated into layers of hard rock. Metamorphic rock, as its name implies, is composed of igneous or sedimentary rock that has undergone basic structural change effected by pressure, changes in temperature or chemical exposure.

For the most part, the valuable minerals found in igneous rock are distributed so sparsely that they cannot be recovered economically; the miner-

Rarely found in its pure state, silver forms many compounds, including the six shown at right, that are rich enough to be mined profitably. The most sought-after of the ores, galena, may contain 10 per cent silver.

STEPHANITE: SILVER ANTIMONY SULFIDE

PROUSTITE: SILVER ARSENIC SULFIDE

HESSITE: SILVER TELLURIDE

ARGENTITE: SILVER SULFIDE

GALENA: LEAD SULFIDE

PYRARGYRITE: SILVER ANTIMONY SULFIDE

als usually crystallize into separate crystals scattered throughout the parent rock. However, certain processes that occur while the rocks are still in the molten stage concentrate some of the elements into commercially workable deposits, or ore. In some cases, the crystals settle through the molten magma and form concentrated deposits at the bottom of the magma chamber. In other instances, in which the metallic minerals are combined with sulfur, crystals separate from the magma, much as lead separates in a smelting furnace, and collect to form large deposits.

As magma cools deep in the earth, the water and other volatile substances separate from the magma and seek to escape. These hydrothermal, or hot water, solutions are the primary transporting agent for the world's precious metals. Because heat always travels toward lower temperature, and volatile fluids toward lower pressure, the hydrothermal solutions rise toward the earth's surface, where both the temperature and pressure are relatively low.

The ore-bearing solutions force their way through natural openings and fissures in the rocks or soak through permeable layers of sedimentary rock. In some instances the hot solutions attack the surrounding rocks chemically, altering their basic structure in a process called contact metamorphism; in other cases the solutions actually absorb the rock they encounter.

Throughout its journey, the magmatic fluid is steadily losing heat. As it works its way upward, it eventually hardens into granite or other igneous rock and, at various stages, may deposit concentrations of ore minerals.

Scientific understanding of where the precious metals come from, and why they are concentrated in certain portions of the globe, has been enhanced greatly in recent decades by the revolutionary new theory of plate tectonics. The theory's basic premise is that the earth's lithosphere — a 90-mile-thick layer of crust and upper mantle — is divided into seven immense moving plates and perhaps 18 smaller ones. The earth's continents ride on these plates like so many rafts frozen to an unseen glacier. The movement of the great plates is exceedingly slow, at most a few inches per year. But over millions of years the cumulative effects of this relentless movement are titanic. The shifting of the plates widens the oceans and narrows them again; it splits continents, then rams them together to raise mighty mountain ranges. The plates spread apart and grow as new crust is formed at midocean ridges. They rub past one another with grinding force, giving rise to earthquakes, and when they collide head-on, one plate may plunge beneath the other in a violent process called subduction.

The significance of plate tectonics for mining geologists is that it provided a new framework for understanding what happens at the subduction zones. Economically, the most important kinds of hydrothermal mineral deposits are sulfide-rich veins in which silver and other valued metals have combined with sulfur to precipitate from the hot-water solution. Gold is not found in the sulfides, but it often occurs near them. A great many of these sulfide deposits have been found along present or former subduction boundaries, where two plates have collided. The reason the sulfide ores are concentrated along converging plates is not completely understood, but geologists theorize that the metals are released from the subducted plate as it slides under the top plate into the earth's interior and melts.

The geology of the Western United States offers an excellent example of the process. The imposing Sierra Nevada formed 77 to 210 million years ago as the Pacific Plate collided with, and plunged beneath, the plate on

Charting the World's Bonanzas

Patterns visible in the distribution of the world's valuable minerals reinforce two of the basic tenets of modern geology. The map below shows the locations of the most important known deposits of gold, silver and platinum and the boundaries of the largest of the moving slabs of crust — the tectonic plates — that make up the earth's surface. Many of the precious-metal deposits are clustered at or near the edges of the major plates and thus offer tangible evidence in support of the basic assumptions of the theory of plate tectonics.

The deposit areas usually are elongated in shape, run parallel to the plate boundaries and are roughly the same age as their host rocks. These characteristics give credence to the hydrothermal theory of the origin of metals, which holds that the energy required to concentrate metallic elements comes from forces re-leased by the collision or separation of the giant plates (pages 74-77).

There are a number of intriguing exceptions to the pattern. Some deposits of great worth, such as those found in eastern Canada and central Russia, are located deep in the interior of large continental land masses, far from any present plate boundaries. Most of these, however, appear in areas where active plate boundaries did exist eons ago, when the planet had a different configuration.

A remaining few, notably the fabled Witwatersrand in South Africa, date from the earliest formative period of the earth's surface and cannot be explained by the theory of plate tectonics. Nevertheless they appear to share with boundary deposits their fiery manner of birth — emanating from molten matter that was driven upward from deep beneath the earth's crust.

On a map of the world's major deposits of noble metals, a significant proportion (*solid symbols*) are found near the borders of six principal plates of the earth's crust. The other deposits (*outlined symbols*) are believed to occur near former plate boundaries, or to have formed before the plates themselves existed.

▲ GOLD
■ SILVER
● PLATINUM

△ GOLD
□ SILVER
○ PLATINUM

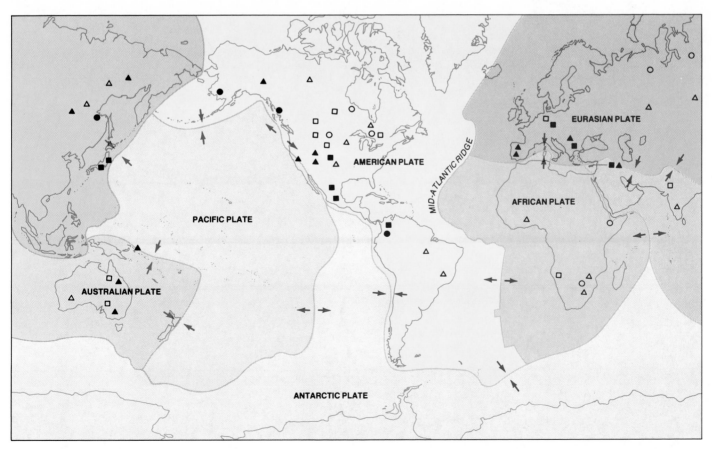

Lumps of pillow lava, mineral-rich basalt exuded along the Galápagos rift in the eastern Pacific, lie at a depth of 10,000 feet. The recently formed sea

floor was photographed by the Woods Hole Oceanographic Institution submarine *Alvin*, whose external arm can be seen in the foreground.

which the American continents ride. Gold, silver and other metals in the ocean floor were released from the Pacific Plate and partly melted as the plate was subducted under the American Plate. These molten magmas gradually rose through the crust of the American Plate, cooled, and accumulated in batholiths — great subsurface masses of granitic rock. In some places the magma broke through the surface in volcanoes — a process that continues today, as evidenced by the violent eruptions of Mount St. Helens since 1978. The metals became concentrated in hot-water solutions emanating from the magma and made their way into cracks and other vulnerable zones surrounding the batholiths. There they formed the extensive veins of gold and silver ore that made up the fabled Mother Lode of California.

Similar tectonic action explains the creation of the mineral-rich Andes mountains along South America's western edge. Drifting westward, South America converged with and overrode the Pacific Plate, which plunged into a deep-sea trench. The Andes were formed by the mighty pressures along the boundary of the convergent plates. Metallic minerals melted from the Pacific Plate as it was subducted. In hydrothermal solutions, the metals ascended through the overlying crustal layers and were finally deposited. In time they resulted in the great veins of gold and silver ore that would enrich the society of the Inca and amaze the Spanish conquistadors.

Fortunately for humanity, the acquisition of much of the world's gold — as well as platinum and a lesser quantity of silver — has required only the simplest technology, or none at all. For millions of years, nature has done the job of releasing precious metal from its voluminous matrix, separating it from baser materials and depositing it in commercially useful concentrations on or near the earth's surface. Such deposits are called placers. Their accidental discovery undoubtedly started primitive people digging in the earth and sifting streams and, more recently, set off the great 19th Century gold rushes to California, Australia, Alaska and the Canadian Klondike.

Placers are created by wind, running water and gravity. The sequence begins when wind, rain and changes in temperature gradually break down the rocks containing primary ore deposits. Given enough time, these forces can wear down entire mountains. Weathering crumbles and dissolves the rock into a collection of loose grains that may range in size from boulders to microscopic particles, liberating the valuable mineral ore in the process. The resulting deposits are known as eluvial placers.

The unique characteristics of platinum and gold make them particularly adaptable to the placer process. Because of their chemical inertness, they cannot be decomposed by the oxygen, humic acid and other chemicals present during weathering.

The carrying away of the weathered matter by natural forces is called erosion. When erosion takes place on a slope, heavier material such as gold, platinum, and infrequently silver, move downhill more slowly than lighter materials, which are washed or blown away. In certain rainless areas, such as the deserts of Australia and Northwest Mexico, wind alone is nature's separating agent. The lighter material is simply blown away and the heavier particles that remain behind are concentrated into deposits called eolian placers (after Aeolus, the Greek god of wind).

But the primary agent in the formation of placers is water, and the deposits it creates are known as alluvial, or stream, placers. When eroded sediment reaches a vigorously flowing stream, the sorting and concentration begin in earnest. It is a complicated, irregular process. Water accentuates the difference in weight between minerals, and moving water sweeps away lighter materials such as quartz while heavier placer minerals such as gold and platinum tend to sink to the bottom. The shape of a particle also influences its rate of settling. A sphere-shaped grain, which has less surface area, will sink more rapidly than a thin, platelike one of the same weight.

The changing velocity of a stream or river is an important factor in how far its load of sediment and heavy minerals will travel. When velocity is doubled, the water's transporting capacity increases about four times. Conversely, when the current slackens, much of the load is dropped.

The trip downstream is not an easy one. The minerals are pounded, scraped and tumbled until the brittle ones among them are reduced to a powder. Even the most durable elements are rounded and compacted. The degree of change offers a clue to how far the minerals have traveled. A prospector who comes upon bits of sharp, angular gold knows that he does not have to look very far upstream in his search for the mother lode.

Alluvial placers occur most often at those points where the flow of a stream suddenly slows down. Deposits are likely to be found immediately upstream from any obstruction that blocks the water's natural progress. They also occur in the potholes and pools that form at the bases of waterfalls and rapids, and just downstream from the confluence of a swiftly flowing tributary into a larger, slower river. Deposits also form in the slack waters along the inner shores of a curve in the stream's course and between the natural ridges formed by ribbed bedrock on the stream's bottom.

Placer deposits are usually gravelly in texture and consist of coarse sand, pebbles and bits of waste rock that may be as large as boulders. Particles of precious metal are typically found mixed with the gravel, although often the heavier minerals work their way down through the gravel and accumulate in crevices in the bedrock below.

While gold is sometimes found in nuggets as big as a man's fist, most placer gold appears in fine specks called dust. The most minute particles, called colors, verge on invisibility. Their presence is revealed only by the yellow cast they give to the surrounding material.

Despite their relative density, some gold and platinum eventually reaches the sea. Here the tide carries much of it away. But the pounding waves and shoreline currents concentrate portions of the gold along rocky coasts and elevated shores as beach placers. In time, the shoreline changes and the gold is buried under sand and other sediment.

When gold was discovered in 1899 on a remote and barren Alaskan beach near the settlement of Nome, frenzied prospectors thought they had stumbled upon the easiest mining of all time. And for a time they had: The black sands of the beach were thick with gold that could be taken with a shovel. But that bonanza was brief. The bulk of Nome's gold, as only the most persistent of the prospectors discovered, lay in placers deposited on ancient beaches that were as far as three miles inland and were buried under as much as 50 feet of frozen tundra.

The origins of history's greatest gold find, South Africa's fabulous Witwatersrand (or "white-water ridge"), are rooted in the same tectonic processes as those that shaped the western edge of North and South America. But the

The Origins of Witwatersrand's Wealth

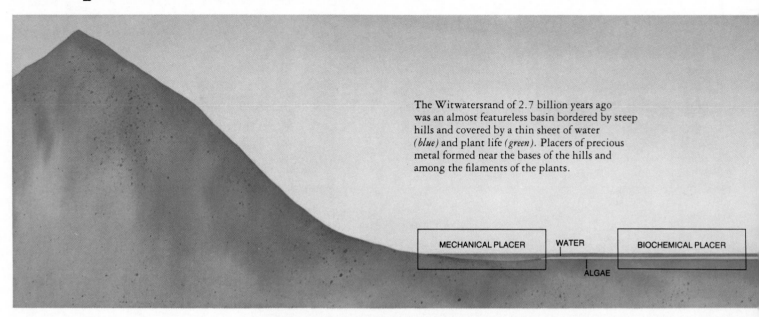

The Witwatersrand of 2.7 billion years ago was an almost featureless basin bordered by steep hills and covered by a thin sheet of water *(blue)* and plant life *(green)*. Placers of precious metal formed near the bases of the hills and among the filaments of the plants.

MECHANICAL PLACER   WATER   BIOCHEMICAL PLACER

ALGAE

Flake of gold recovered from a Rand placer

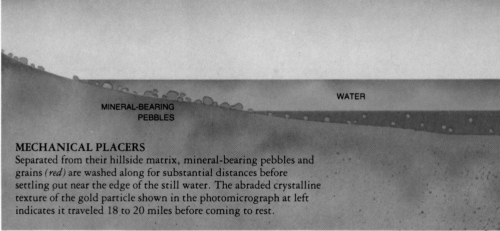

WATER

MINERAL-BEARING
PEBBLES

MECHANICAL PLACERS
Separated from their hillside matrix, mineral-bearing pebbles and grains *(red)* are washed along for substantial distances before settling out near the edge of the still water. The abraded crystalline texture of the gold particle shown in the photomicrograph at left indicates it traveled 18 to 20 miles before coming to rest.

Gold filament taken from carbon fossil

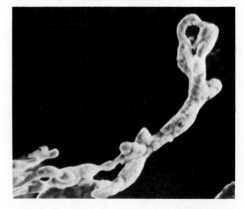

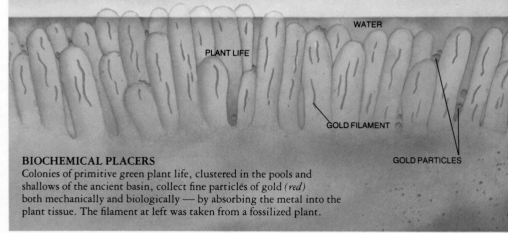

WATER

PLANT LIFE

GOLD FILAMENT

GOLD PARTICLES

BIOCHEMICAL PLACERS
Colonies of primitive green plant life, clustered in the pools and shallows of the ancient basin, collect fine particles of gold *(red)* both mechanically and biologically — by absorbing the metal into the plant tissue. The filament at left was taken from a fossilized plant.

The genesis of South Africa's Rand, the most prolific gold-producing region ever discovered, is thought to have been an intriguing variation on the usual processes of precious-metals concentration. According to geologists, the 150-mile-long crescent of conglomerates rich in gold and uranium owes its existence to a remarkable combination of geological — and biological — events that took place more than two billion years ago.

What is now part of a high plateau was then a watery depression bordered by hills of granite and mineralized lava laced with veins of gold. Over millions of years, mighty winds and torrential rains wore the hills away to nothing and washed the gold into the basin.

Some of the gold particles accumulated in placer deposits by purely mechanical means, sorted by running water. Additional gold was trapped in algae-like plants in the shallow water that covered much of the basin. Some of this metal was further concentrated biochemically when it was absorbed by the plants; the indestructible gold retained its native form through the ages while the plants have survived only as carbon fossils.

Meanwhile, layers of shale and sandstone eroded from the hills. Interspersed with occasional streaks of gold-bearing conglomerate, this material accumulated to a thickness of 25,000 feet. Later, folding of the earth's crust tilted many of the gold deposits sharply downward. Mining these reefs has led to operations more than two miles underground, the deepest mines in the world — so far.

Centuries of storms drive layer upon layer of sediment onto the watery Rand basin. Slender reefs bearing gold and uraninite were separated by thick layers of mud and sand that hardened into quartzite and shale.

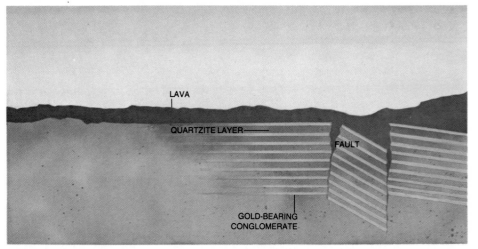

In this contemporary view of Witwatersrand, erosion has leveled the hills and hardened lava covers the basin. Violent movements in the earth's crust have fractured the layers of conglomerate, which dip as much as three miles below the surface. Despite the complications in mining caused by these faults, the Rand yields about 20 million ounces of gold per year.

A 24-sided crystal of pure gold, which measures one eighth of an inch across, displays the symmetry and size that gold crystals attain when they form under ideal conditions of space, temperature and pressure. Such conditions, in which atoms of gold organize themselves into large geometric forms, are extremely rare.

story of how the gold came to accumulate there in the way it did is unique. The Rand, as the ridge is called, is a 300-mile-long crescent of fabulously rich ore. It contains 40 separate mines. Since the first placer nugget was found a century ago, the Rand has yielded 36,000 tons of gold — 31 per cent of all the gold ever produced — as well as substantial quantities of silver and platinum.

Although more than one theory has been put forth to explain the Rand's wealth, geologists working for the South African government believe they have solved the mystery by, in effect, making the ore-laden rocks speak for themselves. Using a scanning electron microscope to supplement the human eye, they have been able to determine, for example, from the distorted shape of some gold particles in the rock that those particles had rolled and tumbled farther from their original source than other particles whose crystalline form remained intact.

From such evidence, the geologists have constructed a picture of what the Rand was like 2.7 billion years ago — when the earth was roughly half its present age. The area was then a great saucer-like basin containing a large, shallow expanse of water, portions of which dried up from time to time. It was a time of violent storms and turbulent winds. Life on earth consisted of microorganisms, including colonies of blue-green algae that grew in the shallow water and puddles of the Rand basin.

The basin was bounded on the north and west by mountains of igneous rock, mostly granite and mineralized lava. Gold was dispersed throughout the rock in countless tiny veins, emplaced by hydrothermal action of volcanic force from below. For millions of years, wind and rain slashed at the mountains, eroding them and washing masses of loosened rock into the basin below. This alluvium contained gold and other minerals, including

Gold that has solidified in a narrow rock fissure is surfaced with thin crystals called trigons. Shown here magnified 100 times, the trigons are evidence that the gold was beginning to form triangular-faced crystals such as the one at left, but the process was halted by the confining rock.

uraninite, the chief source of uranium, along with worthless quartz pebbles.

In the basin, the lightweight gangue — mostly quartz sand and mud — was carried by wind and rain toward the center of the depression. Because it was heavier, the gold stopped moving first — usually at the bases of the mountains or along the shores of the shallow lake. The watery basin functioned something like an immense prospector's pan, with the gold concentrated along its northern and western rim.

Mixed with the heaviest quartz pebbles, the gold settled in layers that ranged in thickness from a few inches to as much as 10 feet. Some of the gold collected in thin layers of algae in the shallow water, trapped as it washed over these living carpets. But in addition, the gold evidently had an atomic affinity for the carbon present in the primitive plants, for it also collected as tiny filaments within the structure of the algae.

The erosion of the gold-bearing mountains into the Witwatersrand basin continued for millions of years. During that time many layers of ore — the South Africans call them reefs — were laid down by sedimentary action. Like occasional golden pages scattered through a thick book, the reefs were separated by broad layers of gangue — mud and sand that eventually hardened into shale and quartzite. In some places the gold layers were only a few feet apart, but sometimes they were separated by half a mile or more of valueless sediment.

At a late stage in the formation of the Witwatersrand, erupting volcanoes gave the basin its final cover. The layers of gold and other minerals, overlain and protected from the elements by sand and mud, were flooded by molten lava. As it cooled, the lava solidified to form the present surface of the Witwatersrand.

When they were first laid down, the Rand's successive layers of gold were roughly horizontal. But over the ages, violent movements of the earth's crust caused the layers to tilt and even to fold. As a result, the gold-rich reefs sometimes appear in outcroppings of rock on the earth's surface, and slant downward for as much as three miles into the ground.

The natural forces that concentrated noble metals at such favored sites as Witwatersrand and the Sierra Nevada have no counterpart in the oceans that cover most of the planet. Yet gold is present in sea water in almost the same proportion as in the earth's crust. Geologists estimate that if all the gold could be extracted from the world's oceans, it would amount to 27 million tons, or one fourth of all the gold thus far mined on land. But the precious metal is dispersed in minute particles, and recovering even a small amount of it requires processing millions of tons of salt water.

Such an enterprise has so far proved unprofitable. Nevertheless, over the past century the challenge of distilling a fortune from the sea has attracted some of the world's most dedicated scientists — and some innovative charlatans as well. In 1897, Prescott Ford Jernigan, a respected Baptist minister in Edgartown, Massachusetts, declared that a way of harvesting gold from the ocean had been revealed to him in a dream. The method called for a zinc-lined wooden box filled with mercury, or quicksilver, to be submerged in the water and left for a full running of the tide. An electrical current was passed through the box, which was called an accumulator, and the gold was supposed to be absorbed by the quicksilver.

Backed by some wealthy parishioners, the Reverend Jernigan built the box and hired a deep-sea diver to submerge and connect the device in Narragansett Bay. The first attempt, certified by government assayers, yielded five dollars' worth of pure gold. More successful tests followed. By the end of the year, Jernigan had founded the Electrolytic Marine Salts Company and had raised $10 million by selling 700,000 shares of stock to enthusiastic investors up and down the East Coast. Only then was it discovered that Jernigan and his diver-partner had been lacing the accumulators with gold. Instead of reaping a golden harvest, the investors lost their millions while the Reverend Jernigan fled beyond their reach to Europe.

The same lure caused Germany's most renowned physical chemist, Fritz Haber, to devote almost a decade of his career to the recovery of marine gold. Haber was a man of checkered achievements. He earned the Nobel Prize for his method of synthesizing nitrates needed to produce the nitrogen fertilizer required in modern agriculture. But he also was responsible for the first modern use of chemical weapons: He advocated and personally supervised the launching of the infamous chlorine-gas attack at Ypres in 1915. Haber's purpose in trying to extract gold from the sea was to enable his defeated nation to pay off its World War I debts. In 1919 he outfitted a ship, the *Meteor,* with a laboratory and a filtration system of his devising.

For 10 years Haber's expedition plied the oceans, testing water from the mouth of the Rhine to San Francisco Bay and the Polar icecap. In the South Atlantic alone, he analyzed 1,635 samples taken from 186 locations. At first the results were promising. Then Haber discovered that tiny traces of gold in the chemicals and apparatus he was using to assay the water were producing false results. Once this was adjusted for, Haber could find no place in the world's oceans where the concentration of gold was worth pursuing. "There is more chance of finding a needle in a haystack than extracting gold from the sea," he wrote, and in 1928 he abandoned the search.

Rich Bounty from a Spreading Sea Floor

The historic Red Sea, once braved by Phoenician traders and the galleys of Solomon and Sheba, has recently become a magnet for seagoing scientific expeditions. Beneath its waters, global theories of sea-floor spreading and the hydrothermal origins of valued metals can be tested in miniature scale.

So far, most of the exploration has focused on an area of about 100 square miles roughly 6,600 feet deep in a rift valley created by the tectonic spreading of the sea floor. Here oceanographers have located deep pools of water that are both unusually warm — up to 140° F. — and eight times saltier than the sea water above them.

These hot brine pools contain solutions of metals in far greater concentrations than normal sea water — in some cases 1,000 times greater. Moreover, the thick layer of sediment underlying the pools is endowed with iron, copper, zinc, silver and gold precipitated from the brine; geologists call the sediment metalliferous mud.

Samples taken of both the brine and the mud offer physical evidence to support the theory that most metal deposits accumulate as a result of a continuing subterranean process (*far right*) in which minerals are extracted from rising magma and the rock around it into hot fluids — in this case, brine. Cooled by exposure to sea water, the minerals precipitate into layers of solid sediment (*near right*) that can be retrieved from the sea, at least in small quantities, by coring machines operated from the surface by the seagoing scientists.

Multicolored layers of sediment evidence the heavy and varied mineral content of a core sample taken from beneath a pool of hot brine in the Red Sea. Gold and silver are most likely to be present in the blue-gray bands of sulfides at top and bottom. The whitish band is calcium sulfate and the reddish brown layer below it comprises a variety of iron oxides.

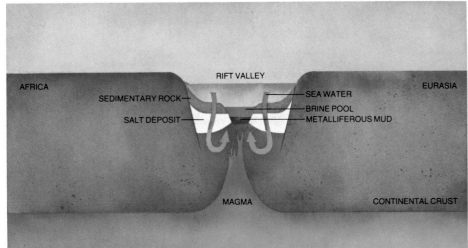

Sea water *(arrows)* penetrates the Red Sea's floor through fissures and faults, warms as it nears the hot magma and begins to rise again. Leaching minerals and salt from the crust, it eventually forms a hot brine pool and a bed of metal-rich mud deep in the central rift.

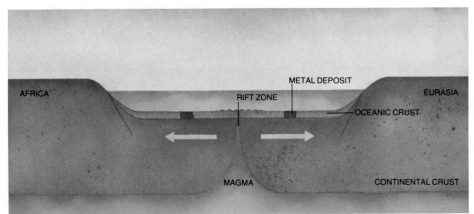

The expanding sea floor *(arrows)* pushes apart the land masses of Africa and Eurasia at a current rate of about 1.5 inches per year. At the same time, the metal deposits in the sea floor are carried steadily away from the rift where they originated.

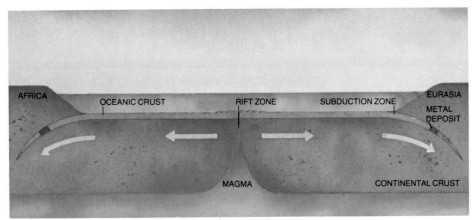

Scientists predict that many millions of years from now the continents will stop moving apart and the still-expanding oceanic crust will be subducted beneath them. The metal deposits will plunge into the earth's viscous interior to be recycled as magma.

Silver crystals synthesized at a refinery display the arborescent, or treelike, shape the metal will assume when it forms under the right conditions in nature. Thousands of microscopic crystals coalesce to form such an aggregate.

Similar frustrations beset those geologists who would extract precious metals from the ocean floors. Where, under the 100 million square miles of ocean, do gold and other metals exist in quantities large enough to merit the awesome undertaking of mining them?

With the variety of coring, drilling and dredging devices developed during the 20th Century, the sampling of the rocks and sediments of the ocean bottom has increased markedly. Yet it is a job barely begun. By one estimate the proportion of ocean floor examined to date adds up to no more than five dredge hauls per million square miles.

Here again, however, the application of plate tectonics offers solid clues to the pursuit of worthwhile mineral deposits. A major premise of tectonic theory is that the plates under the oceans are separating along the mid-ocean ridges. As the plates move apart, the sea floor spreads and new lithosphere is added to each plate along the length of the ridge. The limited evidence so far available indicates that hydrothermal processes are concentrating metals along these midocean ridges. Metallic sulfides, for example, have been found in rocks dredged from the Indian Ocean Ridge. Similar deposits have come from the Mid-Atlantic Ridge, the East Pacific Rise and the Gulf of California. Two other locations under intense study — the Red Sea rift and the mountains of Cyprus — suggest that mineral-rich sulfide deposits occur as consistently along diverging plates as they do in subduction zones.

Cyprus has been famous for its mineral wealth since the days of the Roman Empire, and even earlier. The island's name inspired the Latin word for copper, which has been mined there since 2500 B.C. Cyprus lies close to a diverging plate boundary that runs east and west for the length of the Mediterranean. The island's most important deposits of ore are concentrated high in a volcanic mountain region known as the Troodos massif. After extensive testing, some geologists have concluded that the massif is a slice of oceanic lithosphere, formed as the plates of the sea floor drifted apart, and subsequently thrust upward.

Both the make-up and the layered sequence of the rocks that constitute the Troodos massif match those that lie under the seabed nearby. Moreover, the sulfide deposits that abound in the massif, as well as the layers of sedimentary rock found there, are chemically identical to those found on mid-ocean ridges. The theoretical conclusion is that both the ore and the sediments were formed on the sea floor by hydrothermal processes.

Renowned for its seemingly inexhaustible deposits of copper and iron, the massif is also a valuable source of gold and silver. One of its major mining areas, whose total earthen mass is estimated at 15 million tons, yields one quarter of an ounce of gold per ton and a similar amount of silver. In another part of the massif the sediments underlying a large deposit of sulfur and iron are even richer. They contain more than two ounces of gold and nearly 13 ounces of silver per ton.

The Red Sea, stretching 1,400 miles between the northeast coast of Africa and the Arabian Peninsula, is in the early stage of becoming an ocean. It has been formed by the gradual drifting apart of the major plates bearing the continent of Africa and the land mass of Eurasia. The resultant spreading of the sea bottom along the axis of the Red Sea makes it an ideal laboratory for studying what happens when tectonic plates move apart and new oceanic crust is formed.

Samples taken in 1968 from the bottom of the Red Sea, in an area where the water is more than a mile deep, revealed some of the richest submarine deposits of metallic sulfides ever found (pages 70-71). The deposits were concentrated in three basins near the center of the sea, where the underlying plates have been separating for perhaps 200 million years. In each basin the sulfides are distributed in sediment that lies as much as 330 feet thick. Systematic coring of the top 30 feet of the largest basin brought up layered samples of sediment that was dense with iron, zinc and copper as well as silver and gold. Indeed, geologists put the total mineral value of one 50-square-mile basin at more than two billion dollars.

Overlying the deposits, and saturating them, are hot, salty brines that contain in solution the same metals as those present in the sulfide deposits. From this, geologists deduce that the brines, which bubbled up from vents in the sea floor, are the hydrothermal solutions that give birth to the metallic mineral deposits.

The lessons gleaned from Cyprus and the Red Sea, supported by spot testing elsewhere, have led many geologists to conclude that the 47,000 miles of undersea tectonic boundaries that girdle the globe constitute a road map to the mineral treasures of the oceans. Recovering that treasure, however, remains only a tantalizing prospect, for it will demand an oceanic mining technology far beyond anything that now exists. Ω

THE FIERY ORIGINS OF ORES

There is truth to the adage that nature is the best miner. The gleaming nuggets and veins that constitute the mother lodes of legend are the result of dynamic natural processes that transport minerals to within human reach and concentrate them in deposits worth mining.

Atoms of gold, silver and platinum are scattered throughout the huge reservoirs of molten solution called magma that percolate outward from the searing interior of the planet. Fingers of magma (*below*) probe up through the earth's fis-sured crust, cooling, hardening and precipitating mineral deposits.

The study of plate tectonics — the slow, constant movement of the earth's crust — identifies the rifts at midocean caused by the moving apart of oceanic plates as ideal sites for the formation of mineral deposits. As the plates separate, the earth's crust splits, allowing magma from the interior to ooze upward and form new crust along the rift. The magma cools rapidly and assumes the rounded shape that inspires its name, pillow lava. Metal compounds derived from the hot, upwelling material accumulate on the newly formed ocean floor.

Extracting the minerals from beneath oceans two miles or more deep is a problem that still eludes solution. Fortunately, the undersea deposits are moving steadily toward land (*following pages*). Much of the gold and silver found in the Western United States, for example, is believed to have originated in magma that emerged along the East Pacific Rise, several hundred miles out to sea.

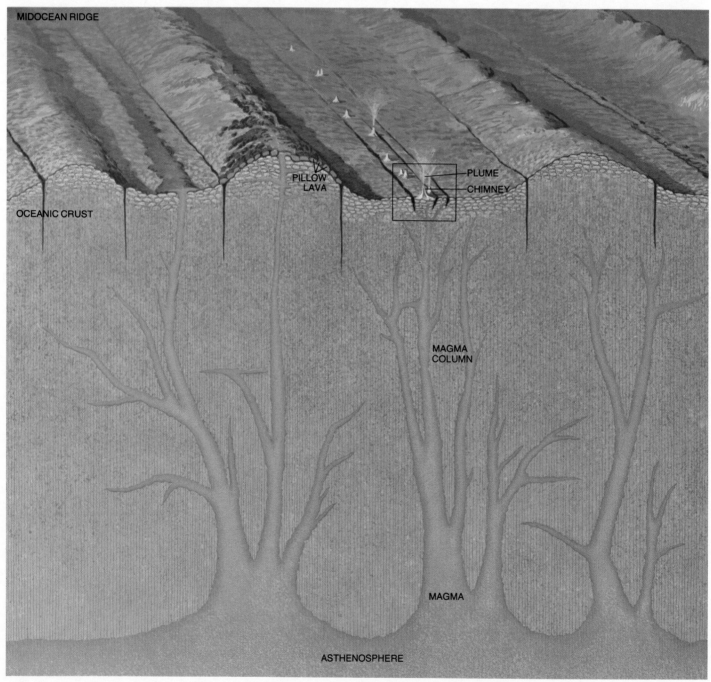

MIDOCEAN RIDGE

PILLOW LAVA

PLUME

CHIMNEY

OCEANIC CRUST

MAGMA COLUMN

MAGMA

ASTHENOSPHERE

Basaltic magma containing traces of gold, silver, platinum and other metals spills onto the ocean floor along a midocean ridge *(below, left)* and congeals into pillow lava. Significant metal deposits are created in the chimneys *(inset),* which emit plumes of mineral-rich water derived from the magma; the process of concentration is shown enlarged below. Near the surface of the ocean floor, cold sea water seeping downward through cracks in the crust mixes with hot fluids released from the magma below. The heated, or hydrothermal, mixture rises, cools rapidly and precipitates concentrated deposits of metallic sulfides and gold in the chimneys.

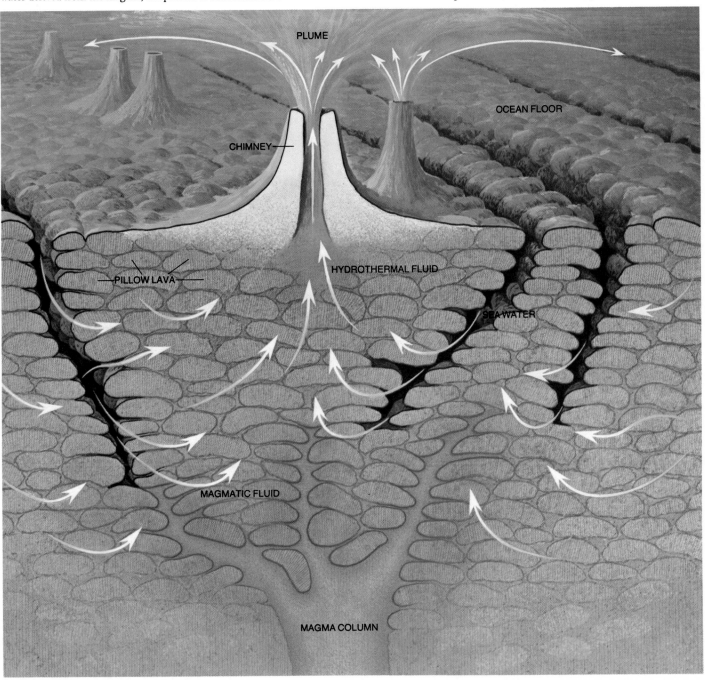

Mixing a Metallic Brew

Inch by inch, over millions of years, oceanic plates with their traces of precious metals advance steadily toward continental plates that in some cases are moving in the opposite direction. If so, the stage may be set for the creation of deposits of precious metals on land.

Where the plates collide, as along the west coasts of North and South America, subduction occurs: The leading edge of the heavier oceanic plate bends into a deep-sea trench and plunges ponderously into the depths of the earth, carrying much of its diffuse mineral load with it. The plate's upper layers, however, are scraped off and partly melted as the plate descends into the asthenosphere — the hot, semiviscous zone below. This molten matter rises through weak spots in the crust of the overriding continental plate. Sometimes it bursts through the surface as an erupting volcano. More often it accumulates in a large underground chamber called a batholith.

Around the edges of the batholith, a hot, metal-rich solution separates from the magma and often combines with cooler ground water soaking down from above. With the heat of the batholith providing the driving force, the hydrothermal solution circulates continuously, leaching additional gold and silver from the batholith and from the surrounding continental rock.

These combined waters rise, cool and crystallize, depositing gold, silver and other metals in existing fractures in the rocks. Often, the result is a metallic vein that eventually may be exposed by erosion, as were the great Mother Lode of California and the prolific Farallón Negro silver fields of Argentina.

The hydrothermal process responsible for most of the world's minable gold and silver deposits occurs near a subduction zone, and above an underground magma chamber, or batholith. Hot, metal-rich fluids derived from the magma and the country rock above it mix with convecting ground water, rise toward the surface, cool and crystallize into veins.

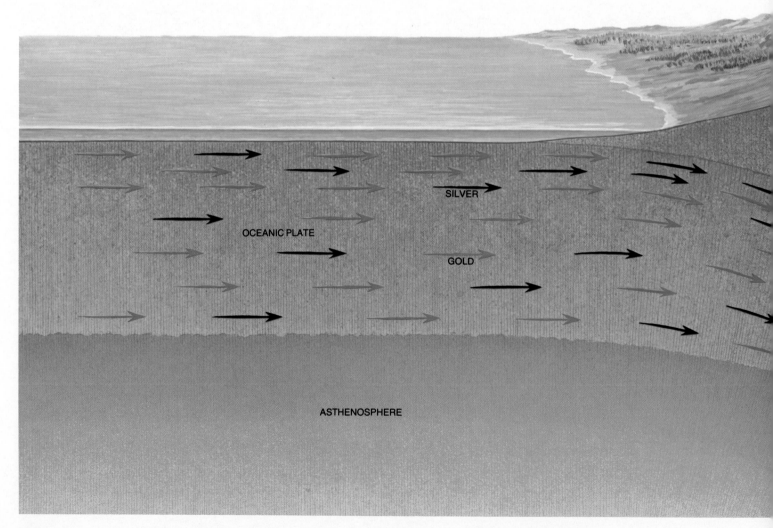

SILVER

OCEANIC PLATE

GOLD

ASTHENOSPHERE

LAVA

VOLCANO

MAGMA COLUMN

CONTINENTAL PLATE

GROUND WATER

HYDROTHERMAL MIXING

SILVER VEIN

GOLD VEIN

MAGMATIC FLUID

BATHOLITH

MELTING ZONE

A Tectonic Source of Platinum

The massive forces that are involved in the movement of the earth's plates are not entirely predictable, and even the aberrations can be beneficial. For example, the kind of plate collision illustrated here, called obduction, is an important source of platinum.

Obduction occurs when the top part of an oceanic plate, instead of sinking into the subduction zone, shears off and is thrust onto a continental plate. The result is a mountain of rock called an ophiolite suite (a term derived from *ophi,* the Greek word for "serpent," be-

cause of the greenish, mottled, snake-like color associated with the rock).

The rock breaks into jagged slabs that lie against one another like toppled dominoes. The hydrothermal process does not occur in obduction, so ophiolite suites are not a rich source of gold and silver. But often they contain platinum-bearing chromite, a tough, chemically inert mineral that erosion *(inset)* later concentrates in minable placer deposits such as those near Krasnoturinsk in the U.S.S.R. and along the Trinity River in California's Klamath Mountains.

An ophiolite suite is created when an oceanic plate collides with a continental plate and shears in two, its upper layer overriding the continental plate. Some of the platinum-rich Ural Mountains were formed in this way as the floor of the ocean that once separated the Russian and Siberian plates was obducted onto the converging continents.

OCEAN

OCEANIC CRUST

OBDUCTION ZONE

OCEANIC PLATE

SUBDUCTION ZONE

ASTHENOSPHERE

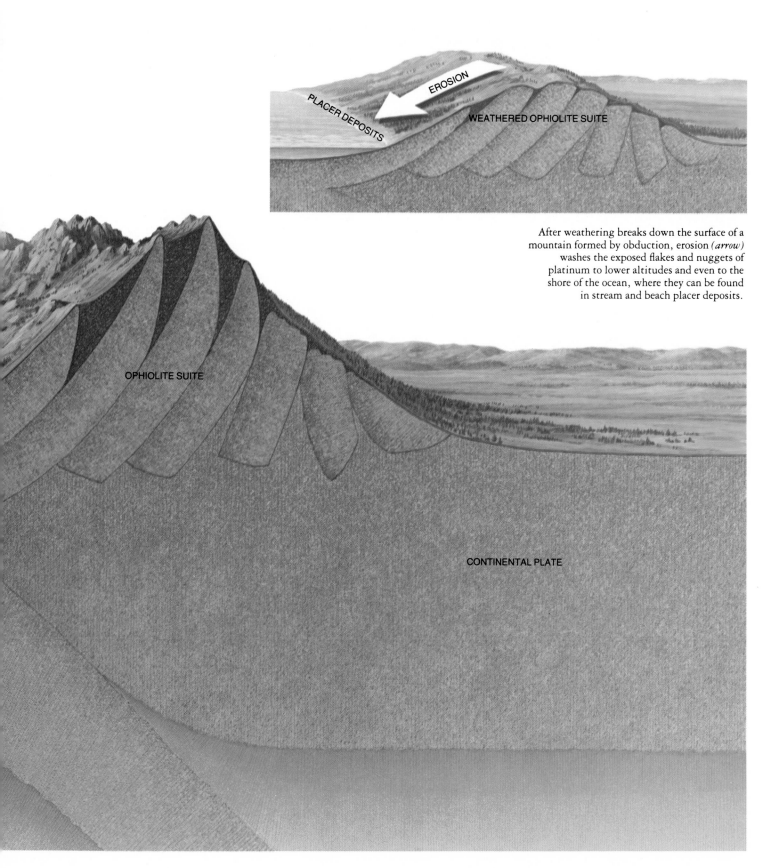

EROSION

PLACER DEPOSITS

WEATHERED OPHIOLITE SUITE

After weathering breaks down the surface of a mountain formed by obduction, erosion *(arrow)* washes the exposed flakes and nuggets of platinum to lower altitudes and even to the shore of the ocean, where they can be found in stream and beach placer deposits.

OPHIOLITE SUITE

CONTINENTAL PLATE

The Making and Breaking of an Ore Chamber

Until 100 years ago, the only known sources of the rare platinum family of metals were placer, or secondary, deposits that had been transported by water and wind. Then the discovery of rich, widely scattered primary deposits — notably at South Africa's Bushveld Complex and at Stillwater, Montana — gave geologists the opportunity to plumb the mystery of the ore's origins. They developed a model of the process that is presented here in three steps.

Metals of the platinum group exist only very sparsely in the crust near the earth's surface but are more prevalent in the mantle below, usually in association with larger concentrations of iron, nickel and copper. Under certain little-understood conditions, portions of the mantle melt, forming mineral-bearing magma that is squeezed upward by pressure within the earth into funnel-shaped intrusions in the upper crust. The intrusions studied so far seem to have formed roughly two billion years ago at the sites of nascent rifts in continental plates. The earth's surface never split, however, and instead of reaching the surface, the magma spread laterally in vast underground chambers where it began to cool and crystallize.

Platinum metals have a strong affinity for sulfur, and in those instances where droplets of molten sulfur compounds formed within the magma, the platinum was concentrated in layers. The process of mountain building later broke up the chambers along fault lines, lifting some of the fractured layers close to the surface where erosion eventually exposed them as outcroppings, ripe for discovery.

In the first stage of the formation of a primary platinum deposit, mineral-bearing magma is injected in spurts through a stemlike feeder into the continental rock above. There the magma expands into a chamber of hot fluid as much as 10 miles deep and an equal distance beneath the surface of the earth.

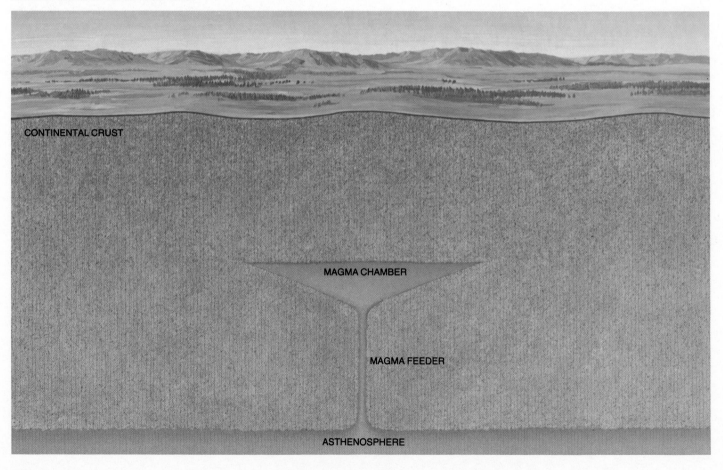

CONTINENTAL CRUST

MAGMA CHAMBER

MAGMA FEEDER

ASTHENOSPHERE

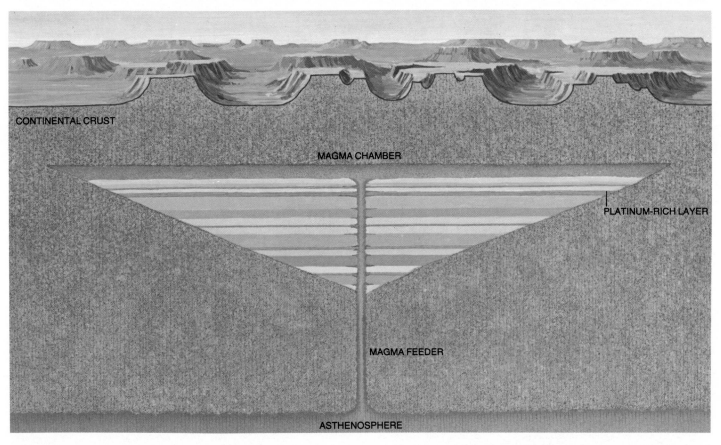

CONTINENTAL CRUST

MAGMA CHAMBER

PLATINUM-RICH LAYER

MAGMA FEEDER

ASTHENOSPHERE

As the underground ocean of magma cools, the minerals in it crystallize in a sequence determined by their atomic composition and arrangement and settle toward the bottom in layers (*above*) based upon their relative density. Sulfides readily extract the platinum (*blue layers*) from the main body of fluid and concentrate it. Forces exerted by the convergence of tectonic plates fracture the hardened strata (*below*). Sections of layered, ore-bearing rock are thrust to the surface, becoming visible clues to the unseen resources below.

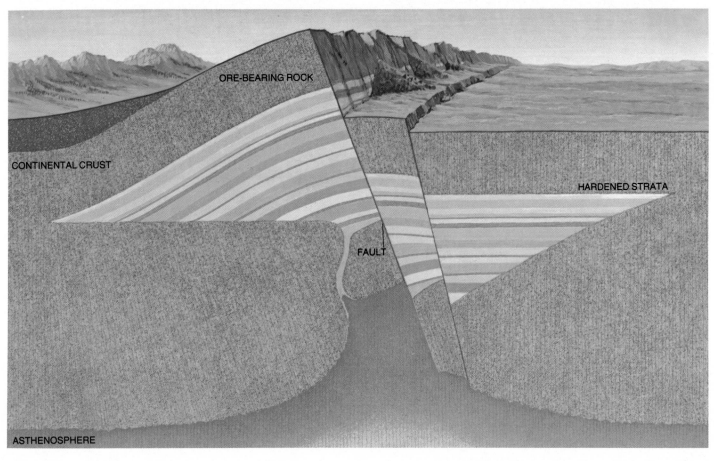

ORE-BEARING ROCK

CONTINENTAL CRUST

HARDENED STRATA

FAULT

ASTHENOSPHERE

81

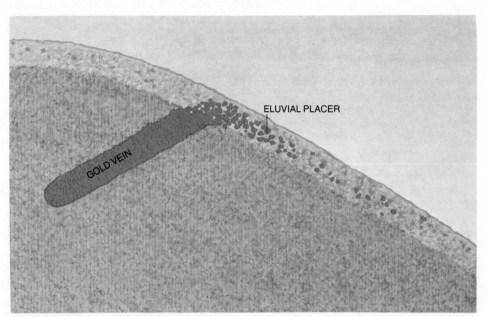

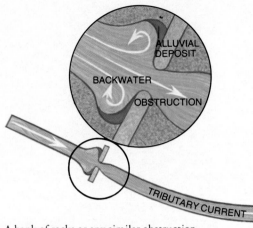

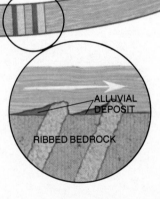

ELUVIAL PLACER

GOLD VEIN

ALLUVIAL DEPOSIT

BACKWATER

OBSTRUCTION

TRIBUTARY CURRENT

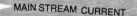

MAIN STREAM CURRENT

ALLUVIAL DEPOSIT

RIBBED BEDROCK

When a vein of gold (*red*) is exposed to weathering on a slope, some of the released particles come to rest a short distance downhill. These eluvial placers, formed without any stream action, often contain large nuggets.

A bank of rocks or any similar obstruction that partially dams the flow of a stream, as shown above, creates a backwater that is a natural site for an alluvial deposit.

One likely place for alluvial, or waterborne, deposits to accumulate is on the upstream side of ribbed bedrock that forms transverse ridges across the bottom of a stream.

A Prospector's Guide to Placer Deposits

The great gold rushes of history — from the Klondike to Australia to Witwatersrand — began when someone spotted a yellow gleam in or near running water. Enough of these rich placer deposits have been found by accident to perpetuate the dream that the uninitiated and uninformed are as likely as anyone to stumble across a mother lode. In fact, an experienced prospector can improve the odds of finding a placer deposit by using basic physics and common sense.

Placers are called poor man's ore deposits because they consist of loose gravels and sand that generally are found at or near the surface and can be mined easily with a minimum of equipment.

By definition, placers are secondary deposits moved by natural forces from their primary source, at a higher altitude. Weathering — the scouring action of wind and rain, and the expansion and contraction caused by temperature changes — frees grains of valuable minerals from the rock matrix (*top left*) of their primary deposit. Eventually, erosion and gravity transport both minerals and rock to running water, which accelerates the separation process while carrying the alluvial material downstream.

A placer concentration can only occur when the valuable matter has a higher density than the waste that is swept away from it and is durable enough to

survive the wearing action of the moving water. Gold and platinum are notable among the handful of minerals that meet these requirements — along with diamonds and some other hard gemstones.

As the drawing on these pages shows, there are several places along a watercourse where the stream is most likely to drop its load of precious metal and allow it to accumulate. All such locations have one common factor — a reduction in the stream's velocity. Thus placer miners, whether equipped with modern, powerful dredges or no more than the old-timer's picks and pans, make their own luck when they look for gold where swift water slows down.

82

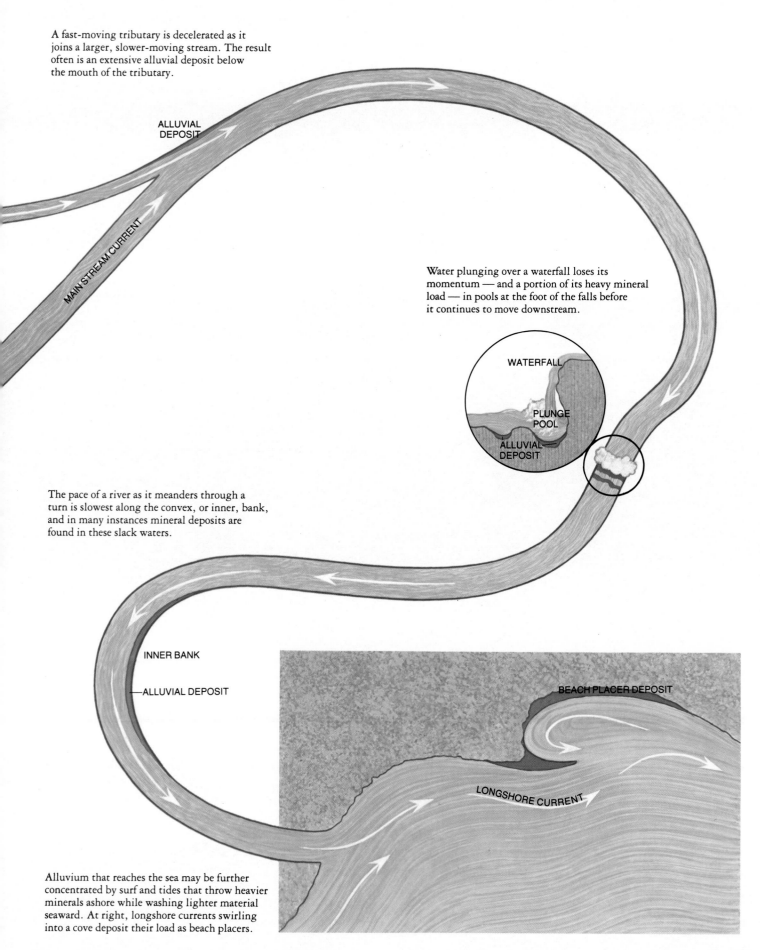

A fast-moving tributary is decelerated as it joins a larger, slower-moving stream. The result often is an extensive alluvial deposit below the mouth of the tributary.

ALLUVIAL
DEPOSIT

MAIN STREAM CURRENT

Water plunging over a waterfall loses its momentum — and a portion of its heavy mineral load — in pools at the foot of the falls before it continues to move downstream.

WATERFALL

PLUNGE
POOL

ALLUVIAL
DEPOSIT

The pace of a river as it meanders through a turn is slowest along the convex, or inner, bank, and in many instances mineral deposits are found in these slack waters.

INNER BANK

ALLUVIAL DEPOSIT

BEACH PLACER DEPOSIT

LONGSHORE CURRENT

Alluvium that reaches the sea may be further concentrated by surf and tides that throw heavier minerals ashore while washing lighter material seaward. At right, longshore currents swirling into a cove deposit their load as beach placers.

Chapter 3

PROSPECTING FOR THE MOTHER LODE

I confess, without shame, that I expected to find masses of silver lying all about the ground. I expected to see it glittering in the sun on the mountain summits." Not yet 30 years old, Samuel Langhorne Clemens, better known to history as Mark Twain, had arrived in 1861 in Virginia City, Nevada, to join the silver boom that followed the discovery there of the fabled Comstock Lode. Like thousands of fellow greenhorns and thousands more veterans of the California gold rush of 1849, the former riverboat pilot and newspaper reporter, with three partners he had met in Carson City, Nevada, staked his claim to a 300-foot-long area he grandly named the Monarch of the Mountain. It would take, he was sure, a week at most to become wealthy; "my fancy was already busy with plans for spending this money."

Twain had given himself over without restraint to the legend of the forty-niners, the quintessentially American faith that any man could strike it rich if only he was sufficiently strong, fearless and lucky. An admiring Twain sought out the veteran forty-niners at work on the Comstock Lode and asked them about their experiences. "They fairly reveled," he reported, "in gold, whiskey, fights and fandangoes and were unspeakably happy."

And so, at first, was Twain. "I crawled about the ground, seizing and examining bits of stone, blowing dust from them or rubbing them on my clothes, and then peering at them with anxious hope." Energized by the irresistible lure of precious metal, he scrabbled over his rugged mountainside. "Of all the experiences of my life, this secret search among the hidden treasures of silver-land was the nearest to unmarred ecstasy." But he found no silver.

It soon became obvious that the men would have to sink a shaft into the mountain to get at the silver, and although Twain's enthusiasm diminished somewhat, he did not give up. They shoveled away until they came to more compact rock, which required crowbars. Then they reached bedrock and had to blast. Two hours of labor with a hand-held rock drill and an eight-pound sledge hammer got them a blasting hole three feet deep and a couple of inches in diameter. "We would put in a charge of powder," recalled Twain, "insert half a yard of fuse, pour in sand and gravel and ram it down, then light the fuse and run. When the explosion came and the rocks and smoke shot into the air, we would go back and find about a bushel of that hard, rebellious quartz jolted out. Nothing more. One week of this satisfied me. I resigned."

Twain tried speculation. With his partners he staked claim after claim and offered them for resale. Optimism bubbled again. "We were stark mad

A weather-beaten prospector, stooped beside the Colorado River, studies his pan for the glint of gold in this 1898 photograph. Lured by tales of sudden wealth, thousands joined the gold and silver rushes of the 19th Century, but for most the only payoff was hard work and misery.

85

with excitement — drunk with happiness — smothered under mountains of prospective wealth." But all the wealth was prospective; "our credit was not good at the grocer's." With hundreds of other prospectors, he was soon reduced to working as a laborer in a Comstock quartz mill for $10 a week. "We had to turn out at six in the morning and keep at it till dark," he wrote. "This mill was a six-stamp affair, driven by steam." Six tall, upright rods of iron, each weighing 600 pounds, "rose and fell, one after the other, in a ponderous dance." His job was to break up masses of silver-bearing rock with a sledge and shovel it into the mill to be pulverized and made into a paste from which the silver could be recovered chemically. After a week of this he resigned and resumed his writing career, as city editor for the local newspaper, the *Territorial Enterprise.*

Twain's experiences, his musings about the fate of the forty-niners and his observations of the increasingly mechanized scene at the Comstock Lode illuminate a fundamental change in the business of mining for gold and silver. The central inventions and applications of this transformation were taking shape in the fierce, brawling competition of the gold and silver rushes of the American West and would soon spread to South Africa's fabulous gold fields. By the end of the 19th Century, the oft-told tale of the lonely prospector stumbling upon the Mother Lode and enjoying eternal wealth would be relegated to the realm of mythology — to join the romantic legends of the Western cowboy and the medieval knight. Recovering precious metals from the earth, unchanged in basic methods for thousands of years, would soon require financial resources, technological know-how and scientific sophistication far beyond the reach of the rugged individualists of the world's frontiers.

Like the forty-niners, the first human prospectors undoubtedly worked placer, or alluvial, deposits from which large flakes or nuggets could be plucked by hand. Where the bits of gold were finer, and scattered through gravel, sand and other alluvial materials, early miners learned to enlist water and gravity; when a container of gold-bearing gravel is washed, the heavy precious metal soon works its way to the bottom and accumulates for easy removal.

The oldest known placer deposits of gold were exploited 4,000 years ago in Egypt. A papyrus document from the Second Millennium B.C. — one of the world's most ancient surviving maps — marks the location of gold mines between the Nile and the Red Sea, an area where tectonic movements of the earth's plates have propelled gold and other metals to the uppermost part of the crust. At about the same time, east of the Red Sea, gold was being dug from open-pit mines.

The ancient Egyptians also discovered gold in the hard rock of certain hillsides in the Nubian Desert and the Sahara. Somehow they had learned that gold was likely to be found amid shiny white quartz, or rocks displaying intense colors such as black, green, blue or rich brown. The brutal labor of tunneling through the rock after gold, sometimes to depths of 300 feet, was the lot of criminals, prisoners of war and even entire families who had displeased the Pharaoh. To be enslaved in the mines amounted to a death sentence. "These miserable creatures," wrote the Greek historian Agatharchides, "always expect their future to be more miserable than even the present, and therefore long for death as far more desirable than life."

The origin of all gold dust, flakes and nuggets is the gold vein, shown at right lacing quartz, gold's usual host rock. Most such deposits are so minute that they are invisible to the naked eye.

Dubbed "frog's hair" by frustrated miners, gold dust is difficult to recover with the traditional techniques of panning and sluicing; the tiny grains easily elude capture.

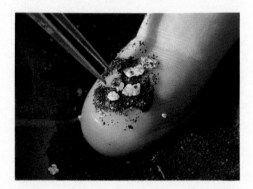

Flat flakes of gold rest in a rich bed of black dirt. Dark soil deposits are an important signal to prospectors; the heavy minerals in the dirt tend to come from the same rock as gold.

This panful of gold nuggets represents one miner's eight-month effort. Visions of large nuggets fill many a miner's dream, but those weighing an ounce or more are rare in reality.

Huge fires were built against the face of the gold-bearing rock and kept burning until the rock was glowing hot. The rock was then doused with cold water. The rapid temperature change cracked the rock, making it somewhat easier for the slaves to break it into manageable pieces with their hammers and wedges. Small boys, who could wriggle through the cramped tunnels more easily than adults, gathered the pieces of rock and carried them to the surface. There, other slaves pounded the rock with iron pestles into still smaller fragments that were then ground in a stone mill until they were reduced to the consistency of fine meal. The meal was scattered on a carved stone table whose top was set at a slight incline. A thin sheet of water trickling gently over the surface washed away the lighter-weight fraction of the rock meal, leaving behind the heavy gold, which was collected with sponges. The gold was then mixed with lead, salt and barley, and fired in a furnace for five days. By the end of this period, the added materials had absorbed impurities from the metal and had vaporized, and a lump of almost pure gold had formed on the bottom of the furnace.

Silver, which is much less likely to be found in placer deposits and is more difficult to separate from rock, came into use in ancient cultures later than gold. Because silver-bearing ore usually contains other metals, most commonly lead, the quest for silver involved not only removing the metals from the ore, but separating the metals from one another. How the ancients learned to do it is unknown, but by the Sixth Century B.C. the Greeks were recovering large quantities of silver by smelting ore mined at Laurium on the southern tip of the Greek peninsula. The mines were controlled by the city-state of Athens, which leased them for operation by private citizens in return for a 4 per cent royalty on the silver produced.

Like the Egyptians before them, the Greeks used slave labor to work their mines. A miner working on a shaft just wide enough to accommodate one person probably excavated about five square yards of rock a month. Hauled to the surface in cowhide sacks, the ore, which contained a great deal of lead, was crushed in mills, sorted and then washed on slanting tables similar to those of the Egyptians. The crushed and washed ore was placed in small furnaces for smelting to separate the lead and silver from any rock fragments. The resulting molten material, called work lead, was fired in a clay crucible. Some of the lead combined with oxygen and vaporized, and the rest of it formed a slag that was absorbed by the porous walls of the crucible. Pure silver was the residue of this process, which was not a particularly efficient one — an estimated one third of the silver wound up in the slag heap. Still, each ton of mined ore yielded about 60 ounces of silver.

The Greeks made other advances in mining technology as well. One problem the Egyptians either had failed to solve or had ignored was the accumulated heat given off by human bodies, the miners' oil lamps and the wood fires lighted to crack the rocks, which could maintain the temperature at an unbearably high level in the mine shafts. To introduce fresh air, the Greeks dug ventilating shafts that intersected with the stopes, or galleries, from which the ore was being removed and installed baffles to direct the air in the desired direction. The Greeks also took some precautions against cave-ins by leaving pillars of rock in place to support the stopes.

One Greek writer called the mines of Laurium "a fountain of silver." He was not exaggerating. All told, the Greeks dug 2,000 shafts during the hundreds of years they worked the mines, and they produced, according to the estimate of a 19th Century Greek mining engineer, 160 million ounces of silver.

For all their success, however, neither the Greek nor the Egyptian miners ever found a satisfactory way to combat the threat of ground water. The deepest shaft at Laurium stopped just above the water table, at a depth of 350 feet. To go deeper was to invite a disastrous flood in the shaft. The problem was first solved by Roman engineers after the Imperial legions had wrested from the Carthaginians the fabulous silver mines of Spain. While honeycombing a 200-mile-long swath of Andalusia, the Romans put an old machine—the water wheel—to new use. When the rich mines of Río-tinto, Spain, were reopened in the 18th Century, a stepped arrangement of eight pairs of Roman water wheels was discovered. Each pair of wheels had raised water about 12 feet to the next level, where it emptied into a reservoir beneath another pair of wheels. Altogether, the system had lifted water about 100 feet.

The Romans also used amalgamation—a chemical technique for isolating gold and silver from rock fragments and other impurities. It involved mixing finely ground ore with mercury, the only metal that remains in a liquid state at ordinary temperatures. The metal dissolved in the liquid mercury to form amalgam, an alloy, which was then heated to vaporize the mercury, leaving the gold or silver behind.

At its greatest extent, the Roman Empire encompassed precious metal deposits from Spain eastward to Asia Minor (page 20). As mining historians Herbert and Lou Hoover observed, "A map showing the extensions of the Empire coincides in an extraordinary manner with the metal distributions of Europe, Asia and North Africa. Further, the great indentations into the periphery of the Imperial map, though many were rich from an agricultural point of view, had no lure to the Romans because they had no mineral wealth."

One part of Europe that the Romans did not covet was Germany. According to the Roman historian Tacitus, "Heaven has denied gold and silver—shall I say in mercy or in wrath? I would not go so far as to assert that Germany has no lodes of silver and gold. But who has ever prospected for them?" No one, apparently, during the last few centuries of the Roman Empire or during the following half millenium of the Dark Ages, when mining virtually came to a halt in Europe.

But mining was resumed, and made its next important advances, in what is now Germany. In 928, a German nobleman rode out on his horse Rame-lus for a day of hunting in Saxony's Harz mountains. Eventually he dismounted and pursued his quarry on foot, leaving Ramelus to graze. On returning, he was surprised to see glittering specks in a patch of ground the horse had pawed. The nobleman led miners to the spot, and their excavations revealed a vein of silver and lead that varied in thickness from 10 to 100 feet. The mine, which was named Rammelsberg in honor of the prospecting horse, was the first to be developed in what would become the most important silver-producing region of Central Europe.

The principal advances in mining technology during the Middle Ages came from the mines of Saxony. Like the Romans centuries earlier, the Saxons used the water wheel to drain their mines, and they devised far superior methods of ventilation, such as crank-operated fans and large bellows that could force fresh air through tunnels as long as 1,200 feet. The

Saxons, of course, had a greater impetus than the Romans to provide healthy working conditions, since their mines were worked by free men and not by slaves. The skill of the Saxon miners and metallurgists became known throughout Europe, and when the ancient Spanish silver mines were reopened in the early 16th Century, Saxon experts were recruited to run them. Among other things, the Spanish learned to employ the amalgamation process.

It was a development whose implications soon spread far beyond the borders of Spain. At the time, the conquistadors were shipping home to Spain the accumulated gold and silver artifacts of the Indian civilizations of Central and South America. When the artifacts gave out, the conquistadors went to the source and began mining the area's rich deposits of silver and gold, using the subjugated Indians as slaves. A steady stream of precious metals was soon flowing back to the mother country. Mechanical means alone — principally crushing, sorting and washing — were used for separating gold and silver from gangue at New World mines until the 1550s. Then a Spanish merchant named Bartolomé de Medina moved to Mexico, bringing with him the technique of amalgamation. Medina's method, which came to be called the patio process, was originally used on Mexican silver ores and was soon adopted throughout the Spanish colonies in the Western Hemisphere.

The silver ore was crushed and ground by mule power in an arrastra, a circular pit paved with stone. Wooden beams connected to a central rotating post were used to drag large blocks of stone around the arrastra, crushing and grinding the ore as water was trickled over it. Eventually, the mined rock was reduced to a fine mud, which was then spread out on a patio and sprinkled with mercury, salt and copper sulfate. Mules or humans walked back and forth over the ingredients to mix them thoroughly — a process that sometimes took weeks. The mass of material was then mixed with water and agitated in large tubs, and the waste was drained off through holes in the bottom. The amalgam remaining in the tub was heated in a furnace to vaporize the mercury. The silver thus refined was cast in ingots, assayed, or tested for purity, and sent to Madrid.

The amalgamation process did more than make silver recovery more efficient; it also made the operation easier to control. Private operators were granted the right to work colonial mines in exchange for giving the Spanish Crown one fifth of the output. Considering the remoteness of the mines and the volumes of silver involved, the temptation to cheat was undoubtedly great, but the royal authorities had an advantage; they could estimate, with a fair degree of accuracy, how much silver an operator had recovered on the basis of how much mercury he had used. And the government was able to keep track of the mercury supply to Mexico, since the metal was derived from cinnabar ores found in the area around the Spanish city of Almadén. The attempt at a mercury monopoly was not completely successful, however. A black market in the metal developed in tandem with illegal mining and silver smuggling, and as much as 10 per cent of all the silver that came out of Mexican mines during the 300 years of Spanish rule may have been illegally produced.

Even so, the Spanish gained enormous wealth from the precious metal deposits of the New World, which was the major source of the entire world's supply of silver during those three centuries. Visiting Mexico in

1803, the German-born explorer and scientist Alexander von Humboldt estimated that in the 110 years before his arrival in Mexico, 1.2 billion troy ounces of silver had been mined by the Spanish. (Troy ounces, about 10 per cent heavier than avoirdupois ounces, are the standard weight in the precious-metals industries.)

As the Romans had disregarded Germany, so the Spanish were indifferent to their territory, California, in which no substantial quantities of precious metals had been found. When Mexico became an independent country in 1821, the Spanish government ceded California to the new nation. And in February 1848, when Mexico and the United States signed a treaty ending their two-year-long war, California became part of the United States. Just a few days before, James Marshall had found gold in the tailrace of Sutter's Mill and had touched off the great California gold rush.

The gold fever spread across much of the world. As many as 200,000 fortune hunters hurried to California, overland from the Eastern Seaboard through the arid Western wastelands, around Cape Horn, across the Isthmus of Panama, eastward across the Pacific. Each new wave of immigrants headed for the gently rolling hills that lay just to the west of the Sierra Nevada. There, streams originating in the mountains slowed, as the land became less steep, and deposited their freight of gold, which had been freed from Sierra quartz by eons of erosion and weathering. By late 1849, miners were frantically at work along a 200-mile-long band that can be traced on a map today because of towns with such names as Placerville and Oroville.

Some of the lucky first arrivals scooped up as much as $500 worth of gold nuggets, flakes or dust from streams and riverbeds every day. The equipment required to gather such riches was modest — basically a strong back, a pick and shovel, and a shallow bowl or pan in which to sort gold out of gravel, silt and sand. The miner might also have a magnet, which he used

In an 1852 daguerreotype, a team of miners and a woman bringing them lunch pose around a recent technological innovation in the California gold fields. Known as the long tom, the contraption could process hundreds of buckets of dirt in one day, a great improvement over the pan and the rocker.

to collect and discard an iron compound called black sand or magnetite that often sank to the bottom of his pan along with gold. And, if he was very well equipped, he used mercury to separate gold from waste rock.

The miners' work was especially exhausting for those gold-seekers who had been doctors, lawyers, merchants, teachers and clergymen. Unused to hard physical labor, they suffered from back trouble and various muscular ills; everyone was in danger of contracting malaria, dysentery and cholera. Many miners were malnourished, since they lived on beans, bacon, bread and flapjacks, with few fresh vegetables or fruits. Inadequate clothing and bitter winter cold increased the chances of disease or death. Sometimes the mining intruded even on death. One group of forty-niners about to bury a deceased comrade noticed that the dirt dug from the grave appeared to contain gold. The officiating minister quickly made arrangements to hustle the corpse to another spot and proceeded to lead the funeral party in prospecting the new diggings.

Panning for gold was not only a difficult but a slow business, especially after the first, highly concentrated deposits were worked over; a skilled prospector could wash only about 50 panfuls of sand and gravel a day. To increase their productivity, the forty-niners employed a few crude separating devices. One, called a rocker, was brought west by prospectors and miners who had used it in working the small gold deposits of the southern Appalachian Mountains. Made of boards or a hollowed-out log, the boxlike device was roughly 40 inches long, 20 inches wide and four inches high, with one open end. At the other end, a hopper housing a sheet-iron sieve rested atop the sides of the contraption. A series of riffles, boards one or two inches thick, was fixed to the inside bottom of the device, which was mounted on a pair of rockers, like those of a child's cradle. While one man dug, a second loaded earth into the hopper and a third flooded it with

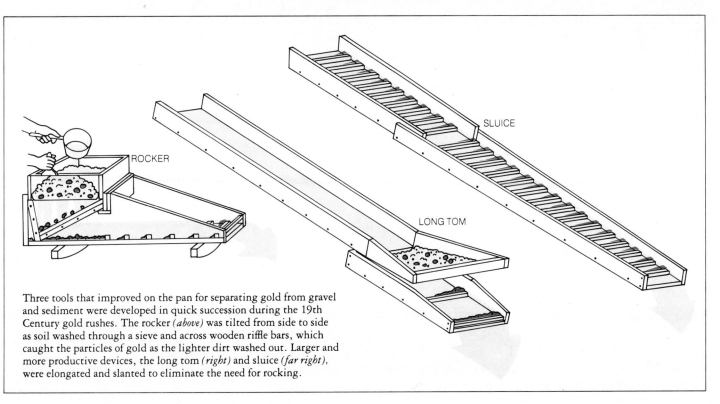

Three tools that improved on the pan for separating gold from gravel and sediment were developed in quick succession during the 19th Century gold rushes. The rocker *(above)* was tilted from side to side as soil washed through a sieve and across wooden riffle bars, which caught the particles of gold as the lighter dirt washed out. Larger and more productive devices, the long tom *(right)* and sluice *(far right)*, were elongated and slanted to eliminate the need for rocking.

An 1855 photograph shows California miners applying the latest in mining technology — hydraulicking. Sixty million gallons of water per day, brought

in from the Sierra 45 miles away by canal and flume (*lower left*), erupted from large nozzles to crumble the rocky hillsides and expose any gold deposits.

water. Gravel and large rock fragments were trapped in the sieve, while the finer bits passed through the holes. The fourth member of the team rocked the machine back and forth to help separate the gold from lighter-weight materials. Most of the gold settled to the bottom and was caught by the riffles when the rocker was tipped from side to side. The waste water drained away through the open end of the device.

Although the men operating a rocker could process much more earth and gravel than they could with pans, they were still limited to about 100 bucketfuls a day. To overcome this problem, the rocker was redesigned and stretched out into a device called the long tom, which could handle four to five times as much earth per day. The long tom was an inclined shallow wooden trough, eight to 25 feet long and 18 inches across at the upper end, widening to perhaps double that at the lower end. The gold-bearing ore was fed into the long tom at the top but, instead of being rocked back and forth, was washed down over the riffles at the end, which trapped the particles of gold from the ore. The long tom was eventually elongated even more, into what became known as a sluice, which was two feet wide and 100 to 1,000 feet long and was constructed in sections on a hillside. Earth was poured in at the top of the sluice, followed by a torrent of water that washed down rocks and gravel and dissolved any sticky clay in which gold might be embedded. Men standing along either side of the sluice shoveled in dirt and lifted out the gangue with a fork. The sluice had a long succession of riffles for snaring the gold.

In 1853, another technological innovation made its appearance in the California gold fields. The idea came to a miner named Edward E. Matteson after he was almost killed in the collapse of the gravel bank of an old, dried-up stream in which he was digging for gold. Matteson decided that it would be not only safer but quicker to use water under pressure to mine such embankments. Called hydraulicking, Matteson's technique entailed diverting a stream from a higher elevation through a flume, or channel, and into a narrow pipe that terminated at the gravel bank. The water, now under considerable pressure because of gravity and successively restricted conduits, passed from the pipe through a narrow canvas hose with an iron swivel nozzle. A single man operating the hose could loosen a ton or more of material per hour and blast the gravel, sand and silt downhill to large sluices. Hydraulicking made it possible to work even the poorest of placer deposits profitably, but its widespread use created other problems; the resulting huge quantities of silt accumulated in rivers, turning them to sluggish, muddy flows. Farmers who required a good water supply for their crops protested so strenuously that, eventually, hydraulicking was severely restricted in California.

During the first few years of the gold rush, mining had required a minimum of investment and had involved the labor of individuals who worked their own claims and kept any profits. But by 1852, the glory days of the forty-niners were fast fading, for the simple reason that the richest placer deposits of the western Sierra slopes had been worked out. At those mines still in operation, it took four men to operate a rocker, as many as six to make use of a long tom and up to 20 to work a sluice. Many a man who had joined the gold rush expecting, like Mark Twain, to find the precious metal "glittering in the sun on the mountain summits" found himself instead laboring for some mining company from sunup to sundown for three dollars

A weary miner surveys the view from a Colorado silver mine. Hard-rock mining became necessary with the depletion of the rich riverbed deposits of the American West in the 1850s.

a day. Those who clung to their independence, scorned working for wages and pressed on into the Sierra Nevada in search of the Mother Lode were able to find, on the average, perhaps $250 worth of gold in a year.

The veins of gold embedded in the quartz outcroppings of the Sierra were called rich man's gold because mining such ores and extracting their metal content was a more complex and expensive business than placer mining. Though hard-rock mines might be discovered and worked initially by small-time prospectors, they were fully developed by men who knew how to organize money, manpower and machines.

The world-famous Comstock Lode that lured Mark Twain to Virginia City was discovered in one mining era and developed in another. The lode was discovered in 1859 by four forty-niners, gypsy prospectors who had moved on from California, from dig to dig, in restless pursuit of the Mother Lode. Henry Comstock, Peter O'Riley, Patrick McLaughlin and Emanuel Penrod named their claim the Ophir, after King Solomon's fabled gold mine. But all they found after weeks of digging was a sticky, bluish black soil flecked here and there with gold. Weary of the backbreaking work with pick and shovel, frustrated by the gluey soil that clogged their sluices and refused to be separated from the meager specks of gold, all the partners sold their shares and moved on.

Comstock was the first to go, delighted to be paid $11,000 for his interest; Penrod was next, selling out for $5,500 and then O'Riley, who commanded $40,000 from a particularly ardent gold-seeker; last was McLaughlin, who could get a mere $3,500 from a young farmer from Missouri named George Hearst, who had to borrow half the purchase price. McLaughlin ended his days as a cook on a ranch in California, O'Riley in a mental institution; Penrod disappeared into obscurity; and Comstock shot himself in despair when he learned the eventual value of his share of the mine — $17 million. George Hearst, on the other hand, was about to become a multimillionaire and the founder of an empire.

Born in 1820, Hearst had come into contact with mining at the age of 15; his family farm supplied meat to workers in a nearby lead mine. Hearst was fascinated by the operation and spent his spare time firing questions at the miners and reading whatever books on mining and geology he could find. He became so knowledgeable that local Indians named him "Boy That Earth Talks To." His father died when Hearst was still a teenager, leaving him to care for the family, three farms and debts of several thousand dollars, in those days a sizable sum. Frugal and industrious, he had paid off the debts within three years; he then headed west to join the California gold rush, leaving his mother, sister and brother to mind the farms.

According to Hearst family biographer W. A. Swanberg, "The gold fever made fools of many poorly equipped young Americans who plunged unthinkingly into adventure and ended up by inhabiting early graves. It did not make a fool of George Hearst. He had the three requisites of the pioneers — strength, courage and ingenuity — and on top of that he had a rudimentary knowledge of mining."

Success did not come quickly. Hearst almost died of cholera during the 2,000-mile trip on horseback to the California gold fields; his first three attempts at placer mining failed; and for a time he opened a general store to avoid working as a laborer in the mines as so many of his colleagues were

Enjoying his growing fame as a codiscoverer of what has been termed the world's largest gold nugget, Bernardt O. Holtermann drapes a proprietary arm over the 630-pound chunk of ore found in Australia in 1872. Although discovered underground and thus not actually a nugget, the large mass of quartz and slate did contain 125 pounds of gold.

forced to do. But during 10 hard years he added to his rudimentary knowledge, and when he inspected the Ophir mine he knew what he was seeing.

The sticky soil Comstock and his partners had despised was in fact argentite, a form of silver ore, and a signal to Hearst that he had bought one of the richest bonanzas of silver and gold on earth. Shortly after taking an option on the claim he made a flat prediction: "You will soon know tons of silver and gold leaving here every month." He had the dark ore loaded on pack mules and carried all the way across the Sierra Nevada to San Francisco, where old mining hands thought he was wasting his time on lead. But when the first 38 tons were processed, they yielded over $80,000 worth of silver, and it was only the beginning.

The veins of silver and gold ore beneath Virginia City were largest near the surface, ranging from 200 to 1,000 feet in width. As the ore descended into the earth, it tapered off in size, and at a depth of 500 feet, averaged about 100 feet in width. Although it was sandwiched between layers of hard, barren country rock, the ore itself was relatively soft and easy for the miners to break up with their picks. But to follow the vein deep into the earth required some means of shoring up the chambers to prevent cave-ins, since the ore lacked the rigidity to support the crushing weight of the overlying rock. The owners of the Ophir called in Philipp Deidesheimer, a mining engineer who had been trained in Saxony.

Deidesheimer designed a framework called the square set — essentially hollow cubes of stout timbers, each of them 18 inches thick and six to eight feet long. As the miners hollowed out the ore, a square set was assembled in the cavity, one unit after another until the mine's interior resembled a gigantic honeycomb. "These timbers were as large as a man's body," Twain wrote, "and the framework stretched upward so far that no eye could pierce to its top through the closing gloom. It was like peering through the cleaned picked ribs and bones of some colossal skeleton. Imagine such a framework two miles long, sixty feet wide and higher than any church spire in America." Mines throughout the American West and Australia as well quickly adopted the square-set system, which was still a standard timbering technique a century later.

The scale of the operations at the Comstock mines, and the enormous amount of capital required to fund them, were unprecedented. In the heyday of the Comstock Lode, 25,000 men worked three shifts a day mining the ore and operating the stamp mills that crushed the ore and amalgamated it in 3,000-pound batches. An endless stream of horse-drawn wagons groaned up the mountain bearing timbers for the square sets, charging what Twain described as "scandalous" freight rates. The costs involved moved Twain to repeat an old Spanish proverb: "It takes a gold mine to run a silver one." But the returns were as enormous as the investment. The two dozen or more mines that tapped Comstock's silver and gold for 40 years produced more than $350 million worth of precious metals; their stockholders struck it rich.

George Hearst used his profits to build a complex empire of mining investments, appearing to gamble boldly but in fact working constantly to minimize his risk by spreading it among several partners — and by applying his growing knowledge of geology. Once Indians had thought the earth talked to him; now he gained a similar reputation among the most experienced miners in the country. It was said of Hearst that he could descend

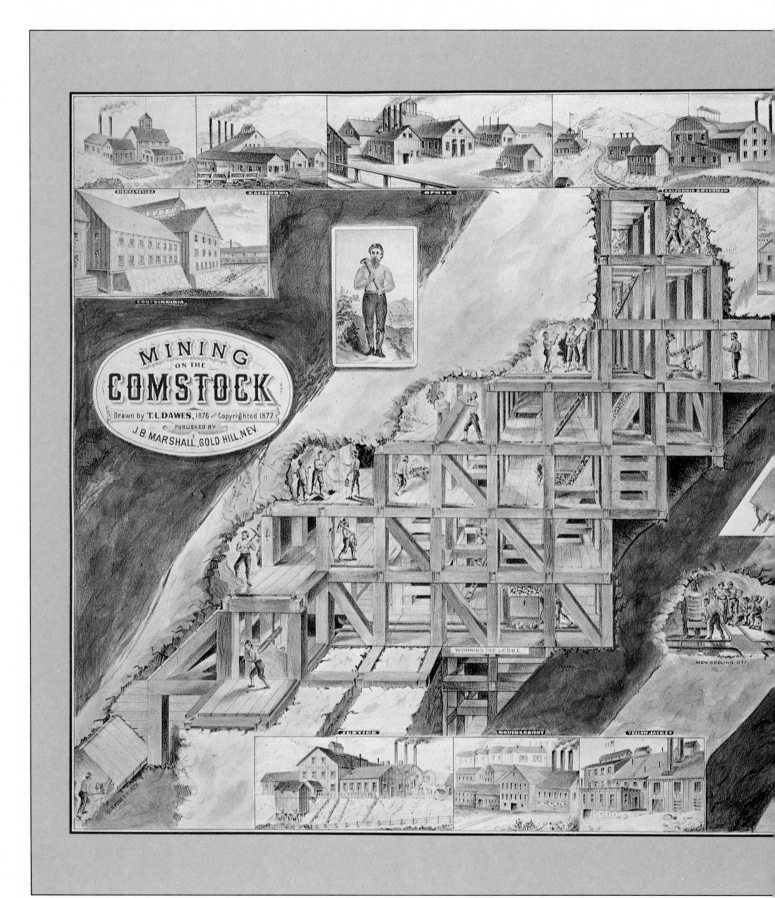

An 1876 lithograph provides details of mining operations at Nevada's fabulous Comstock silver lode at its peak of productivity. Unstable orebodies

Recovering gold from deep in the earth involves extensive effort aboveground as well as below, as this aerial photograph of a South African mining operation attests. Crushed ore is washed in a cyanide solution and moved through a succession of filtration and precipitation tanks to extract the gold, which is then shipped to a refinery for final processing.

rock. He then outlined plans for new stopes that followed closely the contours of the orebody and avoided wherever possible going through country rock. Many members of the Homestake staff did not take kindly to a Harvard geologist telling them that they had been doing their job wrong, and it took some patient persuasion on McLaughlin's part to win them over. But when his scheme for selective mining was put into effect, the value per ton of ore mined more than doubled. Far from closing, the Homestake remains in operation more than a half century later, with an estimated four million ounces of gold reserves yet to be mined. And at mines all over the world, applied geology such as McLaughlin pioneered at Homestake had become crucial not only to the finding of precious metals but to their profitable removal from the earth.

In 1960, for instance, Ralph J. Roberts of the U.S. Geological Survey completed a study of thrust faults in the Tuscarora Mountains of Nevada. The faults had occurred when a sheet of rock, thrust up at a shallow angle, had overridden another rock mass. In some spots in the Tuscaroras, erosion

Sandy hills of tailings — waste from nearly a century of gold mining — add a distinctive feature to the skyline of the South African city of Johannesburg. The mounds, some of which are more than 100 feet high, may one day provide a new source of gold — when a practical method is developed for retrieving the traces of the metal still in them.

106

had created openings called windows in the overriding rock layer through which the underlying rock was exposed. Roberts pointed out in a report that there were metals in these windows — a fact that had escaped notice until then.

The report prompted the Newmont Mining Corporation to take a hard look at the area where prospectors had searched for placer gold for years with little success. Newmont geologists sank more than 800 drill holes to a depth of 80 feet throughout a six-square-mile area. The rock taken from the holes was subjected to atomic absorption analysis, which measures the kind of light waves a mineral absorbs, in order to determine what elements it contains. The analysis revealed that a layer of limestone as little as 10 feet below the surface contained disseminated gold — a scattering of particles so small that most of them are invisible even with an electron microscope and thus had eluded detection by prospectors over the years. Geologists think that hydrothermal solutions transported gold upward through faults and then spread laterally through the limestone, which is relatively porous.

The gold strike in the Tuscarora range has proved to be the most important one in the United States since the Homestake lode was discovered in 1876. Because most of the gold occurs within 300 feet of the surface, the four mines Newmont has put into operation there are huge open pits rather than underground operations like Homestake. After the rock is blasted loose, it is trucked out of the pit to the surface, where it is broken down into gravel-sized pieces. The broken rocks are heaped on asphalt pads and a cyanide solution is poured over them to percolate through the pile. This version of cyanidation, called heap leaching, was developed for low-grade ores that must be processed cheaply to make recovering their gold content profitable. It skips the fine milling used on high-grade ore such as Homestake's, where gold content is high enough to make it worthwhile to build elaborate mills that grind the ore as fine as talcum powder in order to give it maximum exposure to the cyanide solution.

The Newmont mines of Nevada are an important exception, but the rule that guided earlier prospectors still applies: Look for gold where gold has been found in the past. Donald McLaughlin demonstrated the rule when he found more gold at the Homestake, and the lode there now appears to extend as far as 8,000 feet underground. In South Africa, which has produced in the last hundred years about a third of the three billion ounces mined in all of human history, new mines continue to open along the fabulous lode called the Witwatersrand (pages 108-115).

All of these mines are, of course, rich man's hard-rock mines, requiring enormous investments of capital. But poor man's gold, in placer deposits, can still be found here and there. In Alaska, modern prospectors clad in wet suits gather river-bottom gravel and sand with suction dredges and wash out flecks of gold with sluice boxes that differ little from the ones the forty-niners hammered together. In the short Alaskan summer, a diligent and lucky placer miner can find enough gold to support himself comfortably until the return of warm weather. And, in the best tradition of the saloon stories exchanged by the hard-driving forty-niners, placer miners can still strike it very rich: In 1983, a Brazilian prospector working in the Serra Pelada gold field of Brazil dug up a 137-pound nugget that contained a million dollars' worth of gold. Ω

WRESTING GOLD FROM THE NETHER WORLD

Half of the world's annual gold production comes from 41 large mines strung along the 300-mile length of South Africa's Witwatersrand, or Rand. Men have followed the fabulous lode two miles into the earth and still have not reached its end. Some of the Rand's half million workers operate topside mills where ore is processed, but most labor in the thousands of miles of tunnels that have been blasted into the South African rock.

The subterranean world of the miners is dark, hot, humid and dangerous. As shafts are sunk lower, the temperature steadily increases because of heat rising through the rock from deep within the earth. A mile down, the temperature may hover near 100° F., and a mile deeper it rises to more than 125° F. Humidity is often nearly 100 per cent: Water is used lavishly in the mining process and also seeps out of the surrounding rock into tunnels. Although the mines have elaborate air-conditioning and ventilation systems, conditions remain so harsh that new workers must undergo a week of rigorous training in a simulated mine atmosphere to become acclimated and less likely to succumb to heatstroke.

The mines are also noisy. Besides the clatter of pneumatic drills, shovels, rail cars and hoists, the rock itself creaks constantly. Under enormous downward pressure, it can explode without warning in disastrous cave-ins. Electronic sensors monitor the groans of the shifting rock to warn of instability.

The miners of the Rand blast loose and send to the surface every day a quantity of ore equal in bulk to the Empire State Building; this will yield perhaps $22 million worth of gold. But the men who extract this wealth from the earth rarely see even a glimmer of gold — the largest particles must be magnified at least two times to be visible.

In a shaft being sunk into the depths of South Africa's Witwatersrand, miners cluster near a kibble, a receptacle being filled with rock by the hydraulic "cactus grab" suspended at center.

A mucker clears broken rock from a tunnel with a mechanical shovel. The diameter of mine passageways is kept as small as possible to minimize the danger of a cave-in or rock burst.

With no room to stand, miners use their feet to guide a pneumatic drill making holes for dynamite. The drill emits a fine spray that reduces rock dust.

Miners ride a chair lift through a steeply
inclined tunnel connecting two working levels.
In horizontal tunnels, diesel rail cars shuttle
men to and from their work sites.

A haulage-car driver tips a load of ore into a chute emptying onto a central loading platform. From there the ore is hoisted to the surface, where it is sorted and processed.

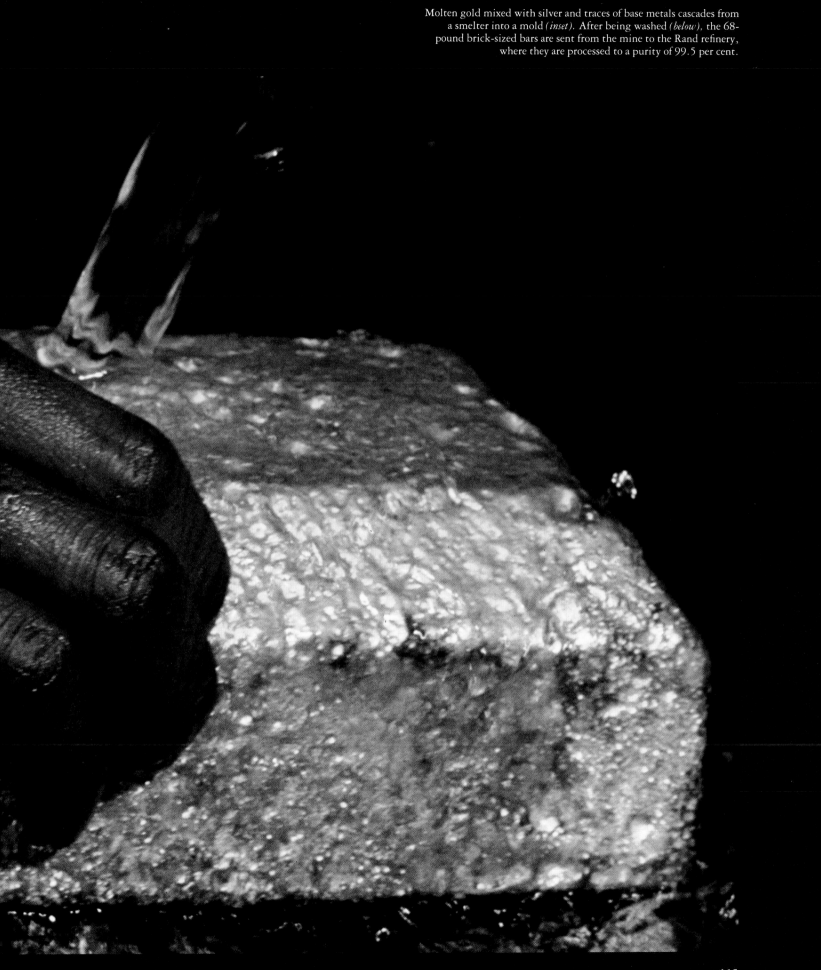

Molten gold mixed with silver and traces of base metals cascades from a smelter into a mold *(inset)*. After being washed *(below)*, the 68-pound brick-sized bars are sent from the mine to the Rand refinery, where they are processed to a purity of 99.5 per cent.

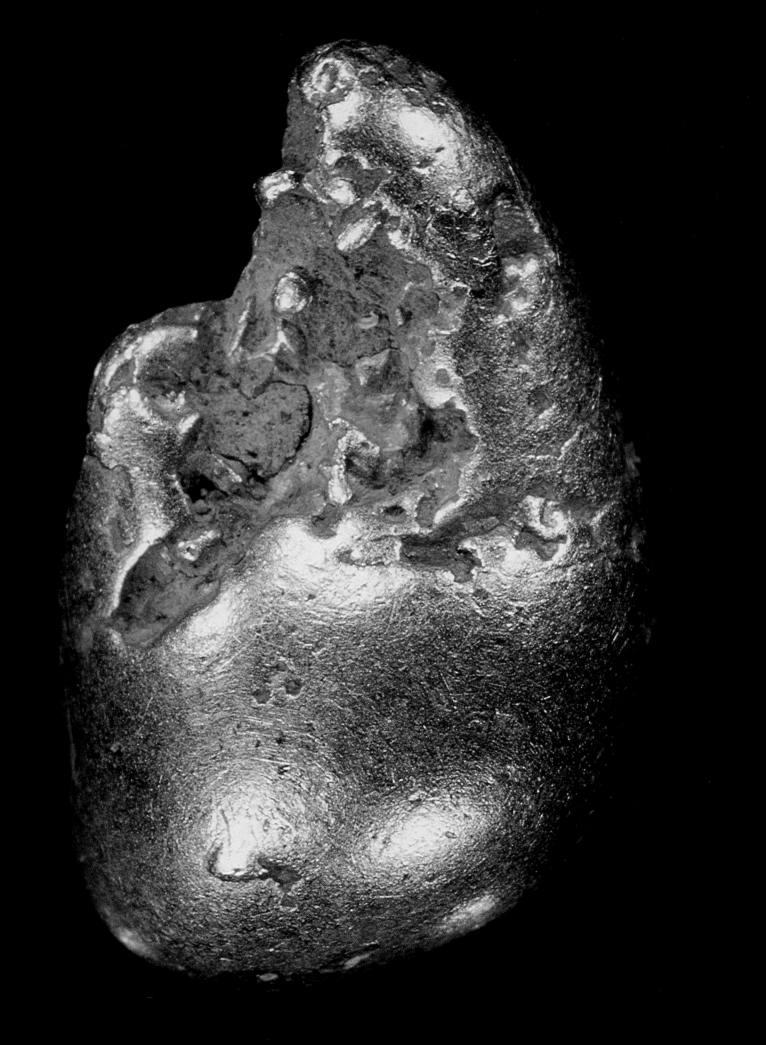

Chapter 4

THE PEERLESS PLATINUM FAMILY

When Howard Hughes wanted a dazzling new actress to illuminate his 1929 movie *Hell's Angels,* he selected from his studio retinue a beautiful 19-year-old extra. She was inexperienced, but she was blessed with star quality and had bleached her hair a stunning snow white. The consummate promoter, Hughes proclaimed that his choice—Jean Harlow—was the very embodiment of Hollywood glamor, and to drive the point home he christened her "the Platinum Blonde."

Associating the actress with the cool elegance of platinum was a public-relations master stroke. Since the mid-1920s, platinum—wondrously rare and five times as expensive as gold—had been gaining increasing favor as an adornment for the throats, wrists, ears and fingers of the wealthy and fashionable. Greta Garbo's platinum cigarette holder bespoke languid elegance. Platinum watches graced the wrists of Scott Joplin and Zelda Fitzgerald; a platinum case held Cole Porter's cigarettes; and Mary Pickford collected virtually anything that was made of platinum. The lustrous, gray-white metal suited the sleek, bold geometry of Art Deco jewelry and served better than gold to complement the glittering fire of diamonds. Indian maharajahs insisted on platinum as the setting for their legendary jewels. In fact, social arbiters had even begun to disparage gold and silver as ostentatious or, even worse, old-fashioned. The Duchess of Windsor, an international style setter, issued a proclamation on the subject: "A fool would know that with tweeds or other daytime clothes one wears gold, and that with evening clothes, one wears platinum."

Because it is much stronger than gold, platinum had for some time been used as a setting for many of the world's finest gemstones. Its growing popularity was credited to the renowned Parisian jeweler Louis Cartier, who introduced the use of the metal as a setting in the 1890s and made it a part of his most exquisite creations for various kings and millionaires. In 1908, the 516-carat Star of Africa was mounted on the British royal scepter in platinum; the Koh-i-noor of India, cut from 800 carats to 106 for the State Crown of England, also rests in platinum.

But the new darling of fashion was destined for a far more important role in industry. Platinum and the five closely related sister metals that make up the platinum family have a broad range of amazing properties. Because of its tensile strength and ductility, one troy ounce of platinum can be spun into a wire more than two miles long. Platinum is all but impervious to acids; it does not tarnish and will not melt except at extraordinary temperatures. Moreover, platinum is a catalyst, one of those rare elements capable

An extremely rare nugget of platinum measures about a half inch in length; far more often, the metal occurs as tiny grains alloyed with other members of the platinum family. Historically, platinum's scarcity and industrial value have driven its price far above that of gold.

of triggering a chemical reaction in another substance while remaining unchanged by the process.

During the 20th Century, these unique properties would make platinum an essential material in such fields as the refining of myriad petroleum products; the making of synthetic fertilizer vital to much of the world's agriculture; the manufacture of textiles, laboratory utensils and contact points for electrical connections; and the practice of medicine and dentistry. Platinum's ascendancy would be swift: Unlike silver and gold, the metal was virtually unknown outside a remote South American valley until the 18th Century. Even after gold prospectors discovered its heavy metal grains in riverbed deposits in the Chocó region of what is now Colombia, they could find no use or market for platinum and considered it a nuisance. Eventually, platinum would ignite scientific curiosity, but it would take years of experimentation to unlock the secret of its family of metals and find uses for their singular properties.

Platinum appears in a few ancient decorative pieces, but no written description of the metal survives, and it may simply have been mistaken for silver. A small strip of hammered platinum set into a gold and silver box from Thebes is believed to have been fashioned in the Seventh Century B.C., and flecks of platinum appear in Egyptian gold pieces dating back to 1400 B.C. During the first three centuries of the Christian era, South American Indians native to the Chocó region became the first people to mine and work platinum, creating nose rings, bracelets, pendants and earrings with a smelting process that Europeans would not duplicate until 1,500 years later.

The Indians first ground the platinum, which is extremely hard even in tiny grains or nuggets, into powder. Then they mixed it with gold dust and heated the mixture on a piece of charcoal over a fire, intensified with a forced draft from a blowpipe, until the gold melted and coated the platinum particles. The mixture was lightly hammered while still hot; then the process was repeated until the gold was diffused throughout the platinum. Radiocarbon dating of artifacts found with examples of pre-Columbian jewelry indicates that the jewelry was made prior to 400 A.D.; by the Ninth Century, the process had been abandoned.

Spanish conquistadors settled in the region in 1690 and immediately began to sift the sands of the rivers for the plentiful deposits of gold to be found there. To the annoyance of the panners, small grains of gray-white metal continually settled out with the gold. The Spanish referred to the tiny pellets as *platina*, a derogatory diminutive meaning "little silver," and spent considerable time trying to separate them from their gold. In 1748, according to the report of a visiting Spanish Navy officer, "Several of the mines have been abandoned on account of the platina, a substance of such resistance that when struck on an anvil of steel, it is not easy to be separated; nor is it calcinable, so that the metal enclosed within this obdurate body could only be extracted with infinite labor and charge."

The platinum had originated, with the gold, among the rocks of the Andes, a mountain range rich in metals and minerals. Erosion had exposed the minerals along the coastal face of the mountains, ground them into particles and washed them into rivers as far south as Ecuador. Gold prospectors finding the troublesome whitish grains would throw them on the ground or, following the lead of local Indians, would toss the metal back into the river where they hoped it would "ripen" into gold. Years later, this casual method of disposal would create havoc when scavengers attempted to recover the discarded platinum. Treasure seekers reportedly leveled the mining town of Quibdó in the Chocó region in their search for platinum nuggets among the foundations of buildings.

In time, New World platinum reached Europe and began to attract the attention of scientists. Sir Charles Wood, a British metallurgist, brought a sample from South America to England in 1741 in hopes that thorough analysis might suggest a commercial use for the metal. Among the scientists who examined the sample was Sir William Watson, a physicist known to history for research into electrical phenomena. Watson subjected the platinum to a number of tests and eventually described the new metal to the Royal Society in 1750. He did not have much to report. His attempts to melt the metal had failed, and as for potential applications he could only

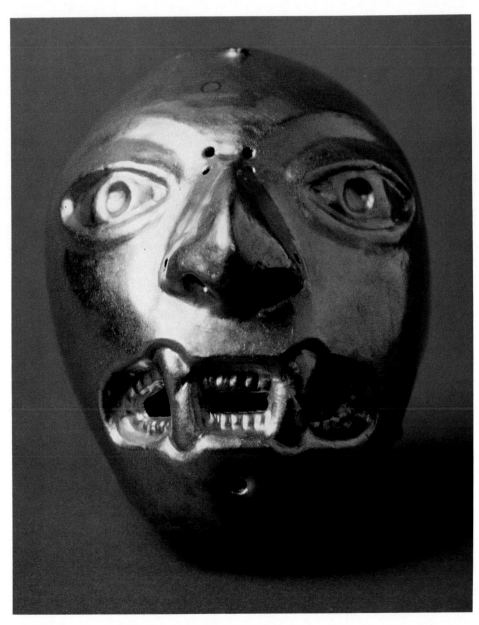

More than 11 centuries old, a mask of platinum-sheathed gold evidences the remarkable skills developed by the pre-Columbian Indians of Ecuador in working the intractable but plentiful platinum that dotted their riverbeds.

postulate that because the metal took a high polish, it might eventually find use in the mirrors of telescopes. Others who examined the seemingly unworkable metal suggested that, at best, it could be used as shot or packed into small bags for clock weights.

Yet platinum was already being put to one highly profitable use — by Spanish counterfeiters. Since 1730, in fact, gilded coins made of the heavy metal had been passed as pure gold, and by 1763 colonial escudos and doubloons and even ingots had been struck from platinum and coated with gold by nefarious characters working in New World mints. Later, even dies from Spanish mints in Madrid and Seville were used to counterfeit coins. Hoping to stem the practice, Spain banned the export of platinum and ordered bogus coins thrown into the sea. When one group of Dutch traders found that they had been defrauded with platinum coins they tracked down the perpetrators and hanged them from the yardarm of their vessel. Nonetheless, platinum counterfeiting continued for many years.

Platinum had provided nothing but trouble for Spain, yet King Charles III was sufficiently intrigued with the new metal to launch his own scientific search for a suitable use. He ordered platinum collected from dumps around New World mints, shipped to Spain and donated to scientific societies throughout Europe. He even offered a generous retainer to anyone who would study the metal.

A French physician, Pierre Macquer, and an apothecary named Antoine Baumé were among those who responded to the offer. In 1757, the two men began trying to melt a small quantity of platinum because, as Macquer wrote, it "is impossible to examine the essential properties of a metal, such as its ductility and hardness, without melting it alone." They placed their sample of platinum — which, like all such ore from the New World, would have consisted of a conglomerate of platinum, iron and sand — in a furnace used for firing porcelain. After five days in the furnace, which was capable of producing temperatures of about 1,400° F. — as Macquer noted, "the greatest degree of fire known" — the crucible holding the platinum had collapsed, but the platinum remained unchanged. The scientists then tried heating the metal with a mercury-coated 22-inch concave burning mirror, a device resembling a huge magnifying glass, which focused the sun's rays to generate exceptional heat. The intense concentration of solar radiation apparently exceeded the heat of the porcelain furnace, for it forced from the ore a few glistening drops of pure platinum, which Macquer and Baumé found easily malleable.

They next exposed the metal to chemical action. Like gold, platinum proved to be soluble in aqua regia, a mixture of nitric and hydrochloric acids. The chemical breakdown produced a solid precipitate, which the scientists then dissolved in ammonium chloride. Inexplicably, each time the experiment was repeated, the precipitate was of a different color, ranging from red to orange to yellow. The chemists were stumped by this inconsistency; the puzzling rainbow would not be explained for nearly a half century.

Satisfied that platinum was as indestructible as gold or silver, Baumé and Macquer hoped to find a way to work it. An experiment conducted by Baumé in 1773 provided a breakthrough that proved platinum could be welded, a necessary step for any large-scale use of the metal. The chemist again heated two ore samples in a porcelain furnace and, perhaps be-

METAL	MELTING POINT (DEG. F)
Gold	1,944
Silver	1,762
Platinum	3,216
Ruthenium	4,190
Rhodium	3,542
Palladium	2,831
Osmium	4,892
Iridium	4,424

The extremely high melting temperatures of platinum-group metals — compared here with those of silver and gold — are the result of the metals' strong atomic bonds. Osmium and iridium, the heaviest of all metals, also have the highest melting points.

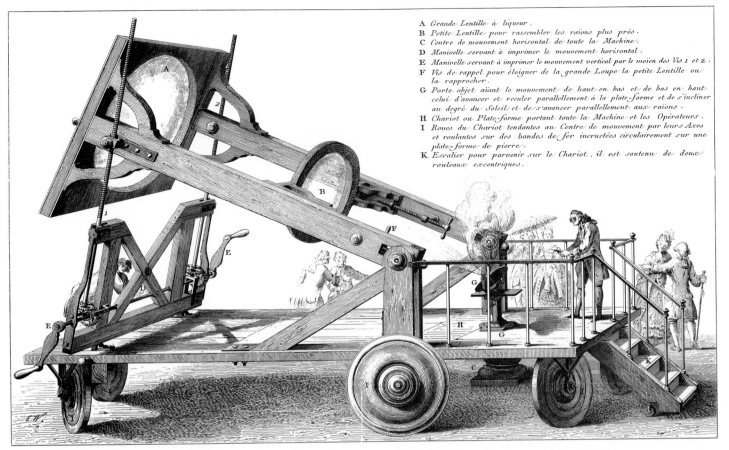

A Grande Lentille à liqueur.
B Petite Lentille pour rassembler les raïons plus près.
C Centre de mouvement horisontal de toute la Machine.
D Manivelle servant à imprimer le mouvement horisontal.
E Manivelle servant à imprimer le mouvement vertical par le moïen des Vis 1 et 2.
F Vis de rappel pour éloigner de la grande Loupe la petite Lentille ou la rapprocher.
G Porte objet aïant le mouvement de haut en bas et de bas en haut celui d'avancer et reculer parallellement à la plate-forme et de s'incliner au degré du Soleil et de s'avancer parallellement aux raïons.
H Chariot ou Plate-forme portant toute la Machine et les Opérateurs.
I Roues du Chariot tendantes au Centre de mouvement par leurs Axes et roulantes sur des bandes de fer incrustées circulairement sur une plate-forme de pierre.
K Escalier pour parvenir sur le Chariot, il est soutenu de deux rouleaux excentriques.

A 1774 engraving depicts the French chemist Antoine Lavoisier attempting unsuccessfully to melt platinum with an enormous burning glass specially built for the purpose. Until his later discovery of oxygen-fed flame, no known heat source could produce the required temperature.

cause of the composition of those particular pieces, was able to obtain a purer platinum than previously produced. Baumé then heated both pieces in a forge and when they were white-hot, he stacked one on the other and struck them with a hammer. He described the result: "They have welded together just as well and as solidly as two pieces of iron. This property of platinum of being malleable when hot and being capable of welding in that state, leads one to hope for the greatest advantages by treating it in this way."

The next challenge was to improve the heating process in order to achieve predictable and thorough melting. To this end, France's Royal Academy of Science directed attempts to melt platinum using enormous burning glasses. In 1774, Antoine Laurent Lavoisier, the father of modern chemistry, obtained the largest burning glass in Paris, a carriage-mounted instrument the size of a small building with a 10-foot focal length and a four-foot-wide lens. The glass had readily liquefied iron, which has a melting point of 2,797° F. When Lavoisier aimed the glass at his grains of platinum, they bubbled and fumed, yet still resisted melting. Nonetheless, it was success enough to fire Lavoisier's curiosity.

During that same year, Joseph Priestley, a Unitarian minister who dabbled in chemistry, isolated oxygen for the first time, a feat that would transform science. Lavoisier immediately began experimenting with the gas and demonstrated that the application of pure oxygen caused a fire to burn hotter. In order to direct a stream of oxygen toward a flame, he designed a complex apparatus consisting of bellows, hoses, scales and weights, faucets, levers and a barrel of water. Using this "gasometer" in April 1782, Lavoisier staged a dramatic experiment in which he propelled oxygen onto a piece of charcoal, hollowed out to hold a chunk of platinum ore, then set it aflame. For the first time in years of laboratory experimentation, the platinum melted completely, producing a round globule of metal. He had at last achieved the melting point of platinum, 3,216° F.

Lavoisier's achievement set the scientific community astir. When the Academy arranged a special meeting two months later in honor of the visiting Grand Duke Paul of Russia, Lavoisier and his gasometer were invited to perform. So on June 6, 1782, the chemist repeated his experiment, this time in the presence of the future Tsar of Russia and his good friend, Benjamin Franklin. Although it made a fine entertainment for scientifically inclined royalty, Lavoisier's technique could produce only very small quantities of metal.

A year later, another Frenchman, Guyton de Morveau, found a way to melt platinum that eliminated the need for the unwieldy gasometer. Chemists had discovered that grains of the metal could be melted at a lower temperature when they were combined with powdered arsenic and potash. Heat caused the arsenic to fuse with the platinum; when heated again, the arsenic would oxidize, leaving only platinum. The problem was that the metal produced this way turned out to be extremely fragile. But by adding powdered charcoal and salt to the initial mixture, de Morveau was able to restore the platinum's characteristic strength, and he used the metal to fashion three crucibles whose imperviousness to chemicals suited them for laboratory work.

The relatively simple chemical technique interested craftsmen and jewelers who could now produce platinum in sufficient quantities to support production of art objects. Marc Étienne Janety, royal goldsmith to Louis XVI, quickly set to work perfecting de Morveau's methods, apparently using platinum smuggled out of South America. Janety succeeded in producing highly malleable platinum from which he fashioned crucibles, snuffboxes, watch chains, mustard pots, buttons, plates, toothpicks and even spoons. Platinum had at last proved both useful and beautiful. It could shine like silver, yet remain impervious to tarnish; its metallic bonding made it stronger than gold. Lavoisier himself displayed one of Janety's platinum coffee pots before the Academy, predicting that "there is no limit to what can be made" if Spain would legalize the export of platinum from its South American colony.

But Spain, suddenly seeing a virtue in the previously despised metal, refused to relinquish its hold on the world's supply of platinum. As interest in the metal continued to increase, a royal edict decreed that "platinum should be worked exclusively for His Majesty as was the case with gold," and in 1778, all subjects who possessed samples were ordered to hand them over without payment. King Charles initiated an attitude toward platinum research that would endure for more than two centuries; he demanded that his scientists work in absolute secrecy to protect their discoveries from duplication.

As a result of that order, the work of Pierre François Chabaneau, a brilliant young French physics professor working at a Spanish university, remained virtually unknown for more than 100 years. Chabaneau sought to produce sizable quantities of malleable platinum using nitric acids and salts, but he was constantly thwarted by the cracking and crumbling of his samples. On one occasion a friend visiting Chabaneau's laboratory found him "engaged in throwing out of the doors and windows his dishes, flasks, and ores as well as all the solutions of platinum which he had prepared with much trouble and difficulty, saying, 'Away with it all! I'll smash the whole business; you shall never again get me to touch the damned metal,' and in fact he broke up all the apparatus of the laboratory."

But three months later Chabaneau at last succeeded, producing a cube of platinum that measured four inches on a side and weighed an astonishing 51 pounds. His sense of humor recovered, Chabaneau casually asked a visitor to move the cube from a table; his friend, not expecting such a heavy weight, could not lift the little cube and accused Chabaneau of fastening it to the table. Obedient to the King's demand, Chabaneau guarded closely the secret of his process (it was discovered in papers more than a century later). He shared it with only one Spanish silversmith, who created for King Charles an exquisite platinum chalice that so delighted the King that he presented it to Pope Pius VI in 1789.

The Napoleonic Wars disrupted research in France and, with Napoleon's first invasion of Spain in 1794, effectively put an end to scientific endeavors there as well. The known techniques for handling platinum were lost for many years. In any case the malleable platinum produced by Janety and Chabaneau, while far more useful than any previously refined, was still too impure and spongy, and too vulnerable to high temperatures, to have any industrial application.

In 1800, a London physician-turned-metallurgist, William Hyde Wollaston, formed a partnership with chemist Smithson Tennant to see if they could plumb the mysteries of platinum's variously colored precipitates and its occasional uncharacteristic brittleness and tendency to burn when fired. Their ultimate goal was to create a commercial-grade platinum. It was to be a fruitful, nine-year collaboration.

Tennant and Wollaston purchased 360 pounds of platinum, unquestionably smuggled out of South America, since Spain continued to forbid its sale. They decided to begin by studying the puzzling black residue that remained when crude platinum was dissolved in aqua regia. Wollaston, working in a laboratory in his home, chanced upon the first breakthrough in 1802. After removing the platinum by dissolving it in aqua regia, he evaporated the excess acid, applied mercuric cyanide to the residue and obtained a yellow precipitate. When he ignited the precipitate, Wollaston was left with traces of an entirely new silvery white metal. He called his discovery palladium after the minor asteroid Pallas, which had been discovered that year.

Palladium proved, like platinum, to be malleable and highly resistant to oxidation and corrosion, but it had a melting point of 2,831° F., much lower than that of platinum, and significantly less density. A pellet of palladium weighing less than a quarter of an ounce could be beaten into a paper-thin sheet 35 feet square. It would later come into use primarily as a chemical catalyst and its alloys as contact points and relays in electronics.

Wollaston suspected the residue merited further investigation and, using a combination of chemicals and acids, he succeeded in isolating yet another metal. He called this new element rhodium, from the Greek word *rhodon,* meaning "rose," which aptly described the color of the solutions of its salts. Of all the platinum metals, rhodium evidences the highest electrical and thermal conductivity. Its great reflectivity and high melting point (3,542° F.) make it an ideal surface for mirrors that are subjected to high temperatures. Extremely rare, and the most expensive

CLASSICS OF A DIFFICULT ART

As chemists unraveled the mysteries of platinum — how to melt it, how to purify it, what uses to put it to — craftsmen kept pace, using the metal in ever more sophisticated decorative art.

The 18th Century French goldsmith Marc Étienne Janety practiced both science and art. In 1788, after a two-year struggle, he perfected a refining method that employed arsenic and yielded larger and purer amounts of platinum than had been possible before. He used this new bounty in masterpieces such as the one shown on the opposite page.

A method of melting platinum powder onto porcelain also was developed in the late 18th Century. At the time, table settings plated with sterling silver were considered the height of elegance. But before long, Spode and Wedgwood factories in England were turning out earthenware sheathed with the equally attractive yet less expensive platinum.

The technological advances of the 20th Century made the metal more readily workable but also created a demand that lofted its price far above that of silver. Henceforward, such utilitarian applications as dinnerware were out of the question, and platinum joined gold and silver as metals of choice for some of the world's finest jewelry.

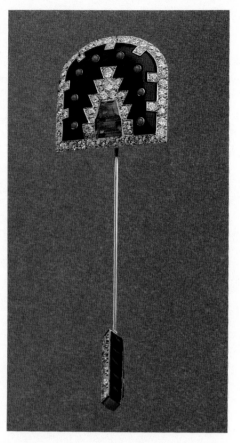

In an 1811 Wedgwood candlestick, platinum applied over ceramic substitutes for the then more expensive sterling silver.

Platinum anchors onyx and diamonds in a pin made in 1920 by Louis Cartier, who stimulated a new interest in platinum jewelry.

An underglaze of blue enamel over earthenware highlights the lustrous platinum that coats a 19th Century British pitcher.

A tea set made by Spode in 1805, trimmed with bands of platinum bordered in gold, marks the earliest use of platinum on porcelain.

This platinum sugar bowl, crafted in 1786, is the only surviving work by Marc Étienne Janety, goldsmith to Louis XVI of France.

of all metals, rhodium typically occurs in proportions of 1 part rhodium to 18 parts platinum.

In 1803, Tennant detected two more new metals following the alternate application of caustic alkali and acid to the platinum residues. He called one of the metals iridium, after Iris, the Greek goddess of the rainbow, for "the striking variety of colors which it gives while dissolving in hydrochloric acid." It was the presence of iridium in crude platinum ore, Tennant realized, that had produced the mystifying rainbow of colored precipitates first noted by Macquer and Baumé. Virtually insoluble in all acids except hydrochloric, and extremely hard, iridium is an ideal alloy to strengthen softer platinum in jewelry. It has a high melting point of $4,424°$ F. but is somewhat brittle alone.

Still, the possibilities of this remarkable ore had not been exhausted. By heating the residue, Tennant was able to extract a volatile, colorless oxide that gave off very strong and unpleasant fumes, later shown to be highly toxic. Tennant named his discovery osmium, from the Greek word meaning "smell." Osmium was found to be the heaviest metal on earth; a cubic foot weighs almost three quarters of a ton. Osmium also has the highest melting point of the platinum metals, $4,892°$ F. Because of its extreme toxicity, which creates problems in storage and refining, there is little commercial demand for osmium; its only significant use is in staining tissues for electron microscopy.

The fifth and last member of the platinum family, and the most difficult to isolate, was discovered 42 years later by the eminent Estonian chemist Karl Karlovitch Klaus, who named the metal ruthenium after Ruthenia, the ancient name for Russia. Not in wide demand, ruthenium, which has a melting point of $4,190°$ F., serves as a catalyst in petroleum refining and, alloyed with platinum and palladium, in electrical contacts. When combined with platinum, it forms one of the hardest metals known, and is used in fine surgical instruments.

Wollaston and Tennant had shown that platinum deposits contain a combination of similar metals, each having distinct properties, which explained why frustrated scientists had found the crude ore so unpredictable. While identification of the new metals assured Wollaston and Tennant a niche in scientific history, their original goal — the commercial production of pure platinum — still eluded them. But Wollaston refused to give up. He embarked on a long series of painstaking experiments requiring subtle adjustments in the application of acids, water and heat, and in the washing and filtering steps. In 1805, after three years of intense work, he arrived at a process that yielded consistently malleable platinum. Shrewd in business as well as science, Wollaston made certain that he would profit from his process by keeping it a secret. His partnership with Tennant ended tragically when, shortly after achieving a final success — the melting of iridium by passing through it an electrical current from an enormous battery — Tennant was killed in a riding accident in 1815.

As Wollaston had foreseen, his platinum bars and ingots proved highly marketable. Among the first products to be made from them were crucibles and lids, balance pans, and blowpipes for laboratory use. The pure metal could be spun into wire $\frac{1}{30,000}$ of an inch in diameter, which was used in the manufacture of high-precision astronomical instruments. But the greatest demand was for the manufacture of flintlock guns; platinum replaced the less durable gold for making the touchholes of guns, where the flint-induced spark ignited the gunpowder. From 1800 to 1821, Wollaston refined 47,000 ounces of crude platinum, most of it smuggled out of South America, producing as by-products 300 ounces of palladium and 250 ounces of rhodium. He tried to market the palladium and rhodium as well, but with little success. The palladium could be used as an alloy with gold in making astronomical instruments and analytical weights, but the only application found for rhodium, which could not be made malleable, was in an alloy with tin used for pen points.

By 1820, commercial demand had seriously depleted the supply of the platinum metals, and Wollaston had to end his industrial sales. Colombia had won independence from Spain, opening up the platinum trade just as the most accessible deposits were giving out after nearly 100 years of panning. On his deathbed in 1828, Wollaston at last revealed the secrets of his technique. By then, new discoveries in the Ural Mountains of Russia had been brought into production; Russia was soon supplying 92 per cent of the world's platinum, a position of dominance it sustained for almost a century.

As Spain had done before, Russia declared a state monopoly, forbidding the export of unrefined metal, limiting all refining to the St. Petersburg mint and permitting mining and sales only with a license, edicts that were flouted rather easily because of the remoteness of the Siberian deposits. Imperial Russia's platinum bonanza fueled the Tsar's political ambitions: The nation had been shut out of the councils of Europe as a result of its contribution to the defeat of Napoleon, and Nicholas I hoped to use his country's growing metallic wealth to enhance Russian power and influence. He invited German and French mining engineers to aid in the development of Russian mineral rights and financed a visit to Russia of the renowned German naturalist, scientist and mineral expert, Alexander von Humboldt.

Humboldt's counsel had been sought by the Spanish three years earlier on the advisability of minting a platinum coin. Citing the lack of a fixed market price for the metal and the fact that platinum did not have the glitter and attractiveness of gold and silver, Humboldt had advised against the idea. For the same reasons, he repeated his counsel to the Russians. But Nicholas rejected the advice and in 1828 issued hundreds of thousands of platinum coins. During the next 18 years, all the problems that Humboldt had foretold—price instability, counterfeiting and lack of demand—plagued the coins, and in 1846 they were finally withdrawn from circulation. Almost a half million troy ounces of the metal had been used in history's only effort to make platinum a medium of exchange comparable to gold and silver.

During the same visit, Humboldt fulfilled a promise he had made to the Tsar; he would not leave Russia, he had vowed, without locating diamonds in the Urals. While studying reports in Russian mining journals on the country's gold and platinum finds, Humboldt recalled that while in South America, he had noted an association among gold, platinum and diamond deposits. While touring the platinum mines of Ekaterinburg, he discussed this relationship with his host, a Russian nobleman who owned several gold fields in the Urals. Trembling with anticipation that he might fall heir to a fortune in diamonds, the nobleman suddenly vanished from the Humboldt party. Hurrying back to his estate, the Russian ordered his overseer to search the sand beds of a gold-bearing river for diamonds. Four days later, a miner appeared with the first diamond, weighing one and a half carats. Within a week, others of even greater size were found. Humboldt did not learn of the discovery until, on his return to the Russian capital of St. Petersburg, a parcel was handed to him containing a diamond. It was the nobleman's token of appreciation.

The failure of its platinum coins forced Tsarist Russia to reconsider its restrictions on the minting and sale of unrefined platinum. Thus in 1850, a young British metallurgist, George Matthey, persuaded one of Russia's major platinum-mine owners to permit the export of a substantial quantity of platinum for refining by his gold-, silver- and platinum-refining firm of Johnson, Matthey & Co.

In 1838, at the age of 13, Matthey had been hired by the precious-metals firm as an apprentice, and in the ensuing years his genius for business, science and metallurgy had forged a link between theoretical scientists and practical metallurgists. The firm — Matthey in particular — had long been interested in platinum, and by securing a reliable supply of placer platinum from Russia, Matthey could direct the profits from the firm's lucrative gold- and silver-refining operations toward the development of a wider use of the platinum metals.

Johnson and Matthey were still processing platinum with the expensive and time-consuming steps pioneered by Wollaston in 1803 and somewhat improved by one of the firm's founders, Thomas Cock. In 1855 Matthey learned of a process developed for melting a newly discovered metal, aluminum, within large blocks of lime hollowed out to serve as crucibles, then fired with coal gas mixed with oxygen. Matthey saw the possibility of a similar application for platinum, but it took 15 years of experimentation before he was able to perfect a lime furnace for use in platinum refining.

Meanwhile, platinum was involved in some revolutionary developments. Michael Faraday obtained some platinum metals, wire and foil from Johnson, Matthey & Co. for his experiments in electrochemistry and magnetism, which paved the way for the development of the electric telegraph. The device required a durable power source, switches that could make and break electrical contact at high rates of speed for long periods of time, and wire that could carry an electric current a great distance. The telegraph patented by Samuel Morse in 1837 used platinum and its alloys for battery electrodes, wire and switch contact points.

In 1863, a 16-year-old night telegraph operator working in a railway station at Stratford Junction, in the Canadian province of Ontario, found some discarded telegraph batteries near the station. An inveterate tinkerer, Thomas Alva Edison stripped the batteries' platinum electrodes, took them home and drew them into very fine wire, which he used in his early attempts to create an electric light. For his early versions, Edison chose alloys of iridium for filaments because it is more dense than the other platinum metals, has a higher melting point and thus better resists the high temperatures produced by an electric current. His tinkering paid off with the development, paralleled by British inventor Joseph Wilson Swan, of a practical incandescent light in 1878. Platinum lead-in wires for the manufacture of light bulbs were eventually replaced by cheaper substitutes, first carbon filaments, then tungsten.

A dredge scours platinum-bearing gravel from 50 feet beneath the Salmon River, near Goodnews Bay, Alaska. The largest platinum-mining operation in the United States, the dredge processes more than a million cubic yards of gravel in the brief, 160-day dredging season, with a yield of up to 15,000 ounces of platinum-group metals annually.

With the growing number of applications for the platinum metals, and the increased production capabilities of Matthey's new lime furnace, the demand for platinum ore began once more to outpace the supply. Had it not been for a fortuitous series of events in Canada, the situation might have become critical.

In 1888, a gold quartz vein being mined near Sudbury, Ontario, suddenly gave out. In its stead lay an entirely different body of ore. As a matter of course, the disgruntled miners sent a sample for a fire assay — and to their amazement discovered they had struck one of the richest copper-nickel deposits yet unearthed in Canada, coupled with lesser amounts of silver and gold. In addition, a small white bead within the ore proved to be platinum. At the time, it was almost unheard of to find so many metals concentrated in one deposit. Even more surprising, the Sudbury strike was the first time that platinum had been found in underground veins rather than placer deposits.

As promising as the discovery was, there was no economically feasible method of separating the platinum metals from the ore that contained copper, nickel, gold and silver. So for several years the ore was mined and the obtainable metals removed, leaving the platinum-bearing residue. Then, in 1924, a process was developed for retrieving the platinum metals from the stockpiled waste by smelting, electrolysis (decomposition accomplished by electric current), dissolution of platinum and palladium in aqua regia and separation of the insoluble rhodium, ruthenium and iridium. Despite its complexity, the process proved to be economical, and Canada rapidly became the world's leading producer of platinum metals. Yet for all its abundance, even the Canadian storehouse was about to be overshadowed.

In 1924, a German consulting geologist working in South Africa, Hans Merensky, was trying to put his life and career back together after having been interned for five years as an enemy alien during World War I. One day, Merensky obtained a small bottle containing grains of metallic ore discovered by a farmer in the Bushveld region. Recognizing that the ore contained platinum, Merensky hurried to the Bushveld. For weeks, the geologist studied the countryside, then made his assessment: A vein of ore containing copper, nickel, iron, platinum and gold ran for 60 miles along a range of mountains in the center of the Transvaal. It would be, wrote Merensky, "the mother reef. If only we can find it."

Merensky's astounding prediction proved to be conservative when the deposit was unearthed later that year. The reef, as South Africans call a mineral-bearing lode, averaged three feet in thickness and extended to even richer deposits near Rustenburg. The vein was found near the surface in some places and as far as 3,000 feet into the earth in others. An enormous mining and refining operation arose at Rustenburg to work the Merensky Reef; by 1975 it employed a labor force of 30,000 in 33 miles of shafts — the longest underground mine in the world. Unlike Canada's Sudbury mines, the Rustenburg mining complex was the first in history to be worked specifically for platinum ores.

Because the location of the orebody is predictable, conventional shaft mining is used. The ore is mined underground by a process used in coal mining known as the longwall system, in which large sections of ore are

Within this 25-foot-high labyrinth of pipes and valves at a New Jersey refining plant, hydrochloric acid and chlorine gas dissolve crude platinum and palladium into chemically workable solutions. Further processing yields chemicals used for making the catalytic converters that remove poisonous nitrogen oxides from automobile exhausts.

extracted in one operation. During refining, the ore is first given a chemical bath in a process known as flotation, which aids in the initial separation of the nickel, copper, iron, gold and platinum ores. The ore is then oxidized into a matte or sludge which is further refined by roasting and leaching with chemicals in order to separate the platinum metals from the nickel and copper matte. The final purification steps for iridium, ruthenium, osmium and rhodium involve as many as 200 different processing stages and can take as long as a year. There are apparently countless variations in the process, details of which are jealously guarded by all modern platinum refiners as trade secrets.

Despite the concentration of platinum along the Merensky Reef and the sophistication of the techniques used to recover it, eight tons of ore must be refined to produce one ounce of platinum. To produce an ounce of gold, by contrast, the South African Rand fields directly south of the Rustenburg mines process only three to four tons of rock.

The Essential Catalyst

Of the myriad industrial uses of platinum, none is more important than its role as a catalyst — an agent that speeds a chemical reaction without itself becoming involved in the reaction. As one example, platinum is vital to the production of nitric acid, 80 per cent of which is used in making fertilizers, the rest in explosives and nylon manufacturing.

To facilitate the production of nitric acid, platinum, often alloyed with its stronger sister metal, rhodium, is woven into gauze to be installed in an oxidation reactor. As ammonia and air pass through the gauze at extremely high temperatures, they combine to form nitrogen dioxide which, dissolved in water, produces nitric acid.

The presence of the platinum accelerates the critical ammonia reaction by as much as a millionfold, thus making mass production of the acid feasible. Almost immune to the ravages of the heat and the chemical changes taking place around it, a pack containing about 50 sheets of the delicate gauze can be used to make 15,000 tons of nitric acid.

But even platinum is not completely impervious; over time, heat causes some atoms of the platinum to be lost, and others to rearrange themselves into crystals on the surface of the wire, gradually weakening the mesh. Most of the lost platinum can be recovered using gauze sheets made of palladium — another platinum-group metal. As a result, the production of nitric acid is one of the most efficient of all chemical processes conducted on an industrial scale.

A sheet of newly made catalytic gauze, magnified 2,000 times by a scanning electron microscope, displays smooth strands of platinum-rhodium wire three thousandths of an inch in diameter, the size of a human hair.

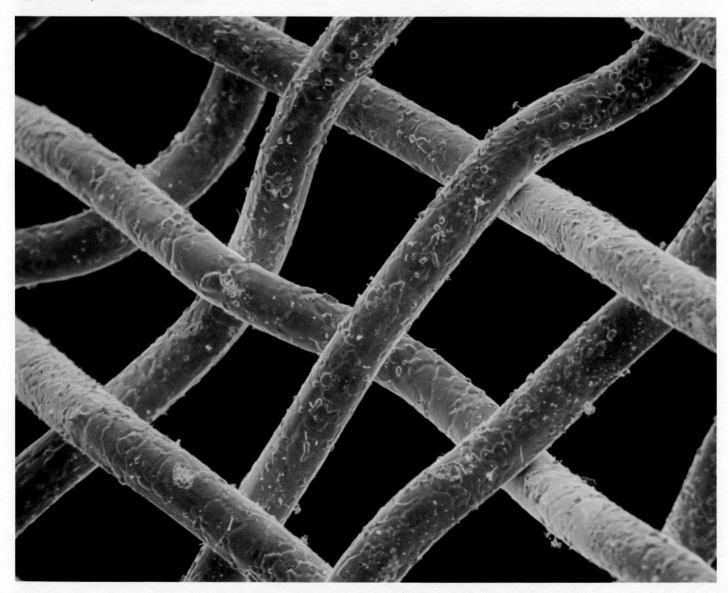

A worker installs a paper-thin sheet of platinum-rhodium gauze in an ammonia oxidation reactor. Oxidation takes place as a stream of heated air and ammonia passes through layers of the catalytic gauze.

After approximately 1,200 hours of use in a nitric acid plant, the gauze shows signs of heavy crystallization, the result of the realignment of surface atoms caused by the high temperatures — 1,620° to 1,740° F. — required by the manufacturing process.

Just three years after the Merensky strike, an Eskimo panning for gold on a river near what is now Goodnews Bay, Alaska, found platinum grains in river gravel. For a number of years thereafter, panning in the area yielded modest amounts of platinum. Then in 1935, a major dredging operation was established to permit the retrieval of platinum-bearing gravel from the floor of the bay, 50 feet below the year-round sheath of ice. Goodnews Bay provides but a small fraction of the world's platinum, equivalent to about 1 per cent of the U.S. annual platinum use. Geological surveys, however, indicate that the region may harbor other commercially viable deposits.

The discovery and development of platinum deposits in Russia, South Africa and Alaska fed the escalating demand for the metal family. Pioneers in the field of photography found that the early silver-based photographic papers deteriorated rapidly. Years of experimentation led to a platinum-treated paper on which delicate shadings could be captured and enduring prints made. Such paper remained the choice for finest-quality photographs until World War I limited the supply of the metal. Platinum also was used in the earliest high-speed internal-combustion engine, developed by Gottlieb Daimler in 1884; air and gas were ignited inside a heat-resistant platinum tube located at the top of each cylinder and heated by a Bunsen burner.

The fact that platinum and its alloys are as impervious to acid as they are to heat made them a key element in the production of the first man-made fiber, later known as rayon, in 1904. The process required that a thick cellulose solution prepared from wood pulp be extruded through tiny holes into a bath of sulfuric acid. A metal device called a spinneret had to be drilled with several precisely placed holes, and it had to be capable of resisting the intensely corrosive solutions involved. The only metal that met all the requirements was platinum, and it is still used to make spinnerets for modern synthetic-fiber production.

But perhaps the most far-reaching utilization of platinum arose from the discovery of its ability to act as a catalyst — a term coined in 1836 to describe the metal's effect on certain compounds. At the time, it was found that fine platinum powder had the ability, when heated, to release, and even ignite, hydrogen from ordinary room air. Further exploration of this phenomenon in the late 1800s led to the successful production of sulfuric acid by oxidizing sulfur dioxide over a catalyst of platinum.

By the turn of the century, a number of scientists predicted that the world's natural nitrate deposits, the source of the nitric acid used to manufacture nitrogenous fertilizer, could not long support the needs of world agriculture. Almost all nitrates were then coming from the saltpeter deposits of Chile. The solution, they said, was to find a method of synthesizing nitric acid in industrial quantities.

However, it was the prospect of war, not hunger, that galvanized German scientists Wilhelm Ostwald and Eberhard Brauer to find that solution. The same Chilean nitrate needed for fertilizer manufacturing is essential to the manufacture of munitions, and in the event of war, shipments of nitrate from Chile might well be cut off. By 1901, Ostwald and Brauer succeeded in creating nitric acid in their laboratory by passing precisely measured quantities of ammonia and air over a catalyst made of platinum at carefully controlled velocities. This basic operation, improved and enlarged, was utilized for commercial production by 1908. But the short life of the catalyst and the relatively large amount of platinum used in the process presented

A newly created crystal destined for use in semiconductor technology emerges from an extremely hot oxide melt. Because of the high temperatures involved, the process requires a crucible made of iridium, which is virtually impervious to corrosion and can withstand temperatures of up to 4,424° F.

Ruby, platinum and chrysoberyl crystals sprout on a two-inch chunk of chemical residue, the extraordinary result of a laboratory accident. During an attempt to create synthetic gems under high temperature and pressure, platinum from the crucible suddenly and unexpectedly crystallized with the gem material.

drawbacks. Karl Kaiser, a German science professor, mitigated the problems by weaving platinum wire into a gauze through which the ammonia and air passed. So successful was this innovation that gauze is still used in catalytic processes.

Platinum was conscripted in World War I both by Germany and the Allies. So successful were the processes for producing nitric acid that neither side suffered from lack of munitions or fertilizer; and all the many other uses of platinum made it a major strategic metal among the combatants. With peace, and a return to prosperity, an unprecedented demand arose for platinum jewelry. Yet platinum's future clearly belonged to industry. In the 1920s and 1930s, platinum contacts could be found in radios, motion-picture cameras, movie projectors and searchlights. By 1934, the glass industry had begun to rely on platinum melting equipment for the manufacture of light bulbs. Airplanes were equipped with high-tension magnetos using platinum contacts. Fountain-pen nibs, thermometers, dental fillings, hypodermic needles and submarine detectors all relied on the exceptional properties of platinum metals. The National Bureau of Standards, as well as comparable foreign institutions, employed weights forged of platinum as the standards by which all else was measured; because the metal would not oxidize, an ounce would be an ounce forever. By midcentury, 200 years of persistent scientific endeavor had transformed a seemingly worthless mineral into an industrial treasure beyond counting. In the years to come, space-age technology would carry platinum and its allied metals still farther, into the distant reaches of the universe. Ω

A growth of palladium crystals, viewed with a scanning electron microscope, displays the dendritic, or branching, structure common to many metals.

Such crystals are formed by passing electricity through a heated solution containing palladium; the current realigns palladium atoms into crystals.

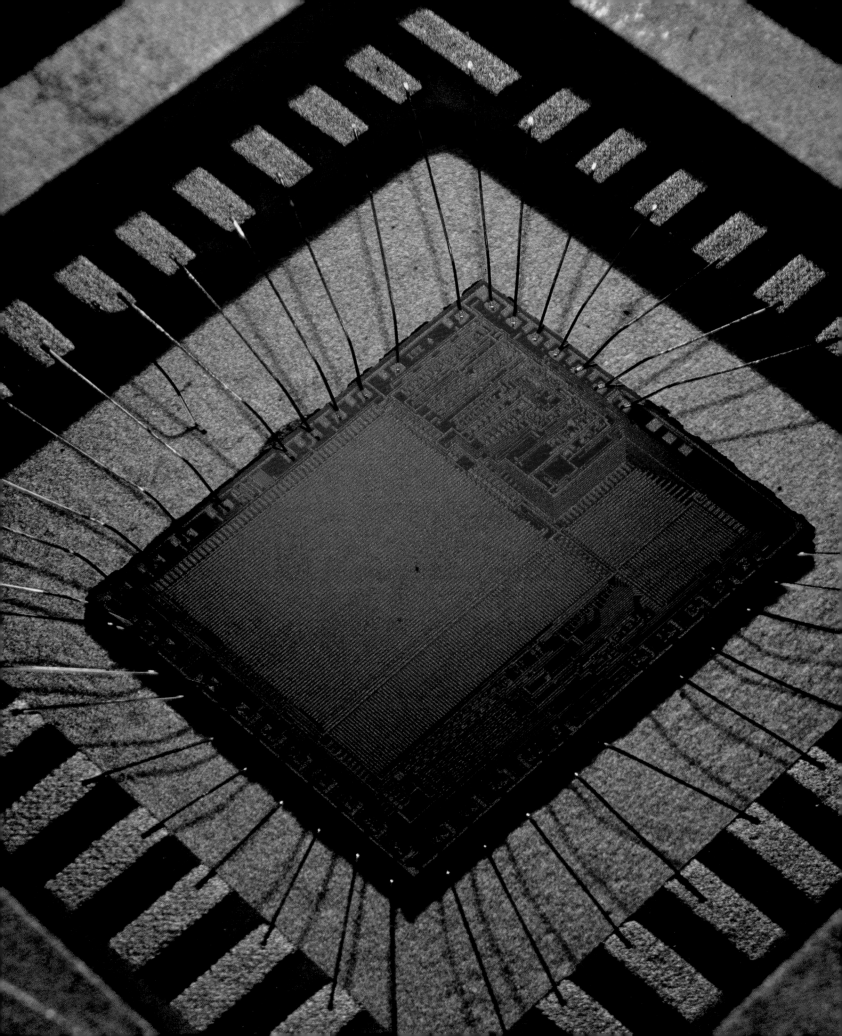

Chapter 5

NEW USES TO SERVE A NEW AGE

t's going to be a bad trip, sir," said a British seaman. "This is Russian gold dripping with blood." The ominous prediction, although based on a wild misconception, proved to be more than accurate. It was made to an officer on the British cruiser H.M.S. *Edinburgh,* docked in the Soviet port of Murmansk in late April 1942, while a series of ammunition boxes were being stacked on the ice-encrusted deck. As everyone on board knew — and as the presence of a formidable guard of Soviet soldiers and British Royal Marines armed with submachine guns confirmed — the boxes contained not ammunition, but more than four million dollars' worth of gold ingots. What the sailor had spotted, and had chosen to interpret as a bad omen, was red paint running down the sides of the just-stenciled boxes, leaving a scarlet trail on the decks as the five tons of gold bullion were being stowed below in the ship's bomb room.

The gold, which had been shipped 1,000 miles by rail from Moscow to Murmansk, was a token payment by the Soviet Union to the United States for what would turn out to be more than $11 billion in World War II aid. The next leg of its journey, aboard the British warship, was to be the 2,000-mile run to Iceland via the Murmansk straits and the Barents Sea — the legendary "Murmansk Run" on which scores of ships and thousands of lives had been lost in the effort to supply the Soviet Army with weapons for its struggle against the invading Germans. The lethal Arctic waters were frequently thick with enemy submarines, surface ships and aircraft.

As daunting as the real and present dangers were, the men on board the *Edinburgh* perceived more ill omens to feed their fears as the convoy they were escorting with 17 other warships steamed single file out of the Murmansk inlet. In addition to the bloodlike paint, there was the number of ships in the convoy — an unlucky 13. And the crew took no comfort from the presence on the bridge, along with their own Captain Hugh Faulkner, of Rear Admiral Stuart Bonham-Carter, who had already had three ships sunk under him. As if these portents were not enough, the ships were sailing at a time of the year when the Arctic daylight has neither beginning nor end, but a dull grayness that would make them targets 24 hours a day. What no one could imagine, for all the fateful omens, was that this vessel and this gold would one day stand as dramatic symbols of the vastly changing role of noble metals in human society.

Late in the afternoon of April 30, while the *Edinburgh* was scouting the waters 20 miles ahead of the convoy, the ship's sonar operator picked up the characteristic echo of a U-boat, dead ahead and closing fast. Captain Faulk-

Gold wires and gold-covered contact points link this integrated circuit — part of a computer-controlled telephone-switching system — with the outside world. Because it is an excellent conductor and is immune to corrosion, gold is essential to electronics.

ner and Admiral Bonham-Carter were on the bridge at the time, and the admiral, claiming that a U-boat producing such a strong signal would be visible to him, told the sonarman to disregard the echo. Off-duty crew members were being piped to afternoon tea when two enemy torpedoes struck the cruiser—one amidships, the other aft. The 613-foot vessel shuddered to a stop, its rudder and two of its four propeller shafts blown away. The *Edinburgh's* radioman signaled the convoy for help, and four destroyers rushed to the stricken ship's aid before the U-boat captain could administer the coup de grâce.

Under tow and escorted by friendly vessels, the *Edinburgh* struggled back toward Murmansk at a speed of scarcely three knots. But the cruiser's oil slick proved her undoing. German aircraft reported her position, and Navy units followed the telltale trail of oil. On May 2, a torpedo fired by a German destroyer reduced the once-proud *Edinburgh* to a listing, shattered hulk. Faulkner gave the order to abandon ship, and as other British ships came alongside to take off the survivors (more than 50 men had died in the attacks) someone asked what was to be done with the four million dollars in gold bullion. According to later reports, the replies were unprintable.

The crippled cruiser refused to sink. In order to deny her precious cargo to the enemy, British warships fired salvo after salvo into the ship, but with little effect. Finally, on the morning of May 2, the British destroyer *Foresight* fired a torpedo into the *Edinburgh* and, as the hushed survivors looked on, the vessel slid toward her Arctic grave 800 feet beneath the Barents Sea, her dead crewmen and boxes of gold bullion still on board. And there she rested for nearly 40 years, seemingly beyond salvage, yet still exuding the eternal lure of gold for those who remembered.

In the 1970s, the discovery of oil beneath the North Sea spurred the development of sophisticated equipment that enabled men to work for long periods in great depths of water. Then, in 1981, a Scottish undersea salvage firm pinpointed the location of the sunken cruiser and sent down a remote-controlled vehicle equipped with lights and television cameras to make a videotape survey of the wreck site. The tapes showed the *Edinburgh* lying on her port side in a nearly horizontal attitude. On the upper, or starboard, side was a gaping torpedo hole. After studying the tapes carefully and consulting the ship's plans, the salvage experts determined that the hole was only about 15 feet from the bomb room that contained the long-lost gold bullion.

After several months of preparation, the treasure seekers were ready to launch the deepest manned undersea salvage operation ever undertaken. Early on September 2, divers on board the 1,400-ton salvage vessel *Stephaniturm* began the long pressurization process that would prepare them for many days of work at depths of 800 feet and more. Then they descended in a diving bell to a spot just 50 feet above the wrecked cruiser.

Linked to the hovering bell by umbilical cables that contained breathing tubes and hot-water hoses that kept the inside of his suit at body temperature, a diver explored the ship and found that the torpedo hole was in fact too far from the gold to be of any use in gaining access to the ship's interior. After laboriously torching a new hole through the *Edinburgh's* heavy steel hull plates, divers finally made their way into the bomb room on September 11. Sifting through thick sediment and debris that included potentially lethal explosives, they began removing the 22-pound ingots.

Of the 465 gold bars that had been stowed away on that bleak day in Murmansk, 431 were retrieved and hauled to the surface. They were worth far more than they had been during the war years, when gold sold on the international market at $35 per ounce. Valued at just under four million dollars in 1942, the salvaged gold was worth $52 million at the 1981 price of $460 per troy ounce. About half of the gold went to the salvage company, two thirds of the remainder was returned to the Soviet Union and the rest to the British government. The United States, having long since been reimbursed by insurance for its losses, received none of the salvaged treasure.

The gold bars that emerged in dripping metal cages from their watery resting place were regarded in a far different light by the world of the 1980s. By the time of their recovery, gold had lost much of its traditional luster as a universal medium of exchange, as had silver, although both metals were still coveted for the same qualities of rarity and immutable beauty that had driven generations of adventurers to seek them in the most distant quarters of the globe. This age-old yearning aside, in the latter decades of the 20th Century, the premier demand for gold and silver came from science and industry, where these noble metals were prized for their astonishingly utilitarian qualities. Silver, for example, had proved to be an essential element in photographic film. Moreover, platinum was fast emerging as a metal with no peer in a variety of technological applications scarcely dreamed of just a few decades before.

The demise of circulating gold and silver coinage in the modern world, and with it the passing of precious-metal backing for paper currency, had actually begun early in the century, during World War I. By the outbreak of war in the summer of 1914, nearly 60 nations — including most of Europe — had adopted the gold standard. But the unfettered circulation of gold among nations and individuals, a basic tenet of the gold standard, quickly became a casualty of the War; enemies engaged in a battle to the death naturally did not exchange gold. Moreover, governments spurred by the fierce demands of combat printed money as fast as they could — regardless of the size of their gold holdings. In all, the Great War cost the participants $202 billion — far more than their combined gold reserves could guarantee. And by 1918, most of the world's nations had determined to maintain those reserves by withdrawing their gold coins from circulation, banning private ownership of the metal and hoarding bullion in central banks.

The United States, late to enter the War, was the last to engage in inflationary wartime financing, and the last to give up the gold standard. It did so in 1917 and simultaneously clamped an embargo on the export of American gold. With peace, however, the United States and several other nations returned for a time to a modified version of the gold standard. Many governments backed their paper money with a combination of gold reserves and American dollars, which had replaced the British pound as the strongest currency in the world. But rampant inflation and a worldwide economic collapse that began with the Wall Street stock market crash of 1929 destroyed confidence in the American dollar. As in troubled times throughout history, people put their faith in gold and rushed to exchange paper currency for the gleaming metal. In the United States, the run on gold became so

Great stacks of gold bricks — each one three and five eighths inches wide, one and three quarters inches high and seven inches long — line the underground vaults of Fort Knox, Kentucky. This main storehouse of United States gold reserves contains more than 300,000 of these bars, each of which is worth more than $16,000 at the official government rate of $42.22 an ounce — a mere fraction of the market value.

frantic that President Franklin D. Roosevelt in 1933 prohibited the export and private ownership of gold coins or bullion; furthermore, citizens were required to turn in their gold and gold-backed bank notes for paper dollars. The price of gold was then set by federal fiat at $35 an ounce. By these methods, and others applied to silver, the U.S. Treasury collected immense amounts of precious metal. A special depository for gold was constructed at Fort Knox, Kentucky, while silver was stored in huge vaults at the United States Military Academy at West Point, New York.

The Fort Knox gold depository was possibly the most impregnable treasure house that human ingenuity had ever devised. Resting on bedrock, the 40-by-60-foot vault was enclosed on its top and sides by 25 inches of steel and concrete. In turn, the vault was ensconced in a granite-and-concrete building with a bombproof roof. Heavily armed guards manned pillboxes at the four corners of this solid fortress, which was also protected by a brigade of motorized cavalry.

The first gold to reach Fort Knox — about 200 tons of bullion shipped by a closely guarded train from the Philadelphia Mint — arrived in January 1937, and much more would follow. Indeed, immediately after World War II, the United States possessed about $23 billion in gold reserves, more than half of the world's monetary supply, and the dollar was again used as backing for the paper money of other nations. But then, in the 1960s, the dollar started to fade as the bulwark of a stable economic order. As other nations gained in economic strength, they began cashing in their dollars for gold. Within a few years the U.S. reserve was reduced by half. The drain was so severe that in 1971 President Richard M. Nixon halted the international exchange of dollars for gold. Then, three years later, the government was impelled by complex considerations of international currency exchange to make an equally far-reaching move with respect to gold, a move that sent lingering shock waves through the world financial community.

At one second past midnight on January 1, 1975, just a heartbeat into the new year, a 12-year-old Michigan girl named Carlenne Brown signed an invoice for a quarter-ounce wafer of gold. She had cashed in two United States savings bonds to raise the $52.79 purchase price. The sale had been arranged by an enterprising coin dealer who enjoyed a brief flurry of publicity for being party to the first such legal transaction in the United States in 41 years. The government, having severed gold from its monetary role, had decreed that American citizens were once again free, as of the first day of 1975, to own, buy and sell gold on the open market.

According to what a number of speculators, investment advisers and metals dealers had been saying — and hoping — during the preceding months, Carlenne Brown's modest purchase should have signaled the start of a veritable gold rush. Millions of Americans, affluent citizens of the richest nation on earth, were expected to join in a mad scramble to buy ingots, bars, wafers and coins of the long-forbidden precious metal. In anticipation of such a stampede, prices on the bellwether London gold market had risen to almost $200 per ounce, an increase of fully 25 per cent in just a few months.

Predictions of gold fever proved to be somewhat premature. To be sure, Americans were intrigued at the prospect of squirreling away a personal gold reserve, and they flooded dealers with inquiries about price and availability. But they did not join in an orgy of gold buying. Within days, the price had slipped below $175, and before the year was out it dipped to less than $130; by 1976, the price of an ounce of gold would fall as low as $103. It almost seemed that the gleaming yellow metal had at long last lost its ancient attraction, that it had become — as the economist John Maynard Keynes had once called it — nothing more than a "barbarous relic" of more credulous times.

Then in 1978, the world was hit with a series of crises, both real and perceived. Vietnamese military forces attacked neighboring Cambodia, threatening to plunge Southeast Asia into war; Iran, its Shah deposed, drifted deeper into anarchy; coups and revolts seemed endemic worldwide; it appeared that Arab oil-producing states might precipitate yet another energy crunch; inflation continued to ravage the U.S. economy. Once again, people began to turn to gold for a financial anchor in uncertain times. By early February 1979, jittery buyers had bid the price of gold to an all-time high of $254 per ounce. As a leading New York gold broker put it, the metal's price had become "a kind of Dow Jones index of investor anxieties." Some gold enthusiasts were predicting that prices might someday top the $300 mark.

In Asia, Latin America and Europe, gold had always been a favored investment. Now, gold fever had struck the United States as well. Of the 54.2 million ounces of gold that changed hands in 1978, more than 20 per cent, or 11.5 million ounces, was sold in America. One of the hottest-selling items in the gold trade was the one-ounce South African coin called a krugerrand: Of the six million krugerrands that were struck in 1978, well over half were snapped up by U.S. buyers. Nicholas Deak, chairman of the nation's largest retailer of the coins, marveled that some of his customers were high-school students. "They just walk in," said Deak, "and say they have a little money and they want to buy a krugerrand."

Throughout 1979, the gold market swelled with buyers, among them oil-rich Arabs and canny Swiss bankers. By early September the London price had hit $340; on one occasion it shot up by $27 in just a few hours. The experts remained convinced, however, that $400 would prove to be an unassailable barrier above which prices would not advance.

The experts were wrong again. Amid mounting world turmoil and the increasingly troubled international economy, gold continued to escalate. In early January of 1980, the price rose by $148 per ounce in a week, to a record $660. One single day's increase of $74.50 was more than twice what the metal's total value had been as recently as 1971. Silver, too, was rising, and hopeful sellers all over the world rushed to market with precious-metal jewelry, tableware and watches. Said one Los Angeles metals buyer: "We have really got a panic here. The stampede is on." In New York, a desperate customer begged a dealer to extract a loose gold tooth; when the dealer declined, the man went on his way but returned a few days later with tooth in hand.

By mid-January, gold had soared past the $800 mark, leaping by a breathtaking 34 per cent in just five trading days. At such heady prices, a single ounce of gold was worth more than 500 pounds of hamburger meat, and the worldwide stampede showed no signs of slackening. Asked to account for gold's wild ride, sociologist and author Neil Smelser called it a classic case of panic. "The people who are dealing in gold," he said, "are operating under the fantasy that the world economic structure is going to collapse. They are living by the myth that the only thing that will survive is gold."

Such dreamers were soon hit hard by harsh reality. Realizing that the metal had been bid up too high and too fast, investors began selling off their holdings, knocking the wind out of the market. More rapidly than it had risen, the price of gold began to plummet. On one heart-stopping day of trading, gold tumbled by $145 an ounce, and the metal was soon flirting with the $400 level. Through the rest of the year gold prices rode up and then down, approaching $700 an ounce before dropping back to the $400 range, where they remained for several years.

Silver prices tumbled as well, much to the chagrin of numerous small-scale speculators who, finding the price of gold beyond their reach, had put their money and their faith in silver. For a time, such faith had seemed justified, the money well spent. At the start of 1979, the metal sold for less

SMELTING SILVER FROM SCRAP

The annual industrial demand for silver constantly exceeds the amount of the metal mined each year; the resulting imbalance of supply and demand provides the impetus for a vigorous secondary industry — silver reclamation.

When the demand for silver drives its price up, supplies of reclaimable silver multiply. When the metal soared to $50 an ounce in 1980, many people rushed to offer their old coins, flatware, jewelry, tea services and the like to refinery furnaces. In addition to the process of scrap refining detailed here and on the following pages, significant yields of silver are derived from used photographic chemicals. The metal can even be wrested from old rags used by jewelers and electronics workers. In all, reclamation makes a considerable contribution to the world's silver supply; about 20 per cent of the silver that enters the market each year has led a previous life in another refined form.

A shipment of scrap metal containing various proportions of silver *(right)* arrives at a metal refinery for weighing and melting in a furnace. Flames destroy all nonmetallic elements, such as the circuit boards in the electronics scrap shown below.

Samples of melted scrap metal are combined with a lead mixture and poured into molds. Droplets of the lead combine with any precious metals present, forming metallic buttons that sink to the bottom and leave the residue — called slag — on top.

Extracted from the slag, the lead buttons are placed in forms called cupels and heated until the lead is oxidized and absorbed into each cupel. The bead of metal remaining is then tested to determine the silver content and thus the value of the shipment being processed.

Shipments with similar silver content are
combined, then melted to remove impurities.
The metal is cast into bars for further refining.

The final step uses an electric current and
chemical solutions to create a sea of crystal silver
that is about 99.95 per cent pure.

Freshly refined silver awaits its new role;
shaped into bars, strips, grains, ingots or wire
(below), the silver is ready for use again.

than seven dollars per ounce; by early September, silver had surpassed even gold's market performance, more than doubling in price. And that was just the beginning. By early January of 1980, silver was selling at five times its price of a year before; on January 17, it hit a historic high of $50.

The biggest and best-known beneficiaries of this steep escalation — and in time the biggest losers — were Nelson Bunker Hunt and W. Herbert Hunt, scions of the late Texas billionaire H. L. Hunt, who had founded the family fortunes when he won his first oil leases in a spirited poker game. His sons had inherited his gambler's instincts. At about the time of the 1973 Middle East War and the subsequent Arab oil embargo, the Hunt brothers began to buy silver futures — contracts to purchase the metal at a specified date and price — as a hedge against inflation. By the time silver reached its January high, the Hunts and some close associates held bullion and futures contracts totaling more than 195 million troy ounces. Sold at the peak price, this awesome hoard would have fetched close to $10 billion.

To be sure, even rumors of such a heavy sale would have driven prices down. But rumors about the Hunts' intentions did not involve attempts to unload their vast holdings. Rather, it was whispered in some quarters that the billionaire brothers were seeking to corner the world silver market, a goal that had last been achieved in 1717 by the mighty Bank of England.

Whatever the Hunts intended, their plans were derailed when the bottom dropped out of the silver market. On a single day in January, the price plunged by $10 per ounce; by late March silver struck a low of just $10.80. And the Hunt brothers, for all their fabled wealth, had to borrow more than a billion dollars to cover their losses. Later, a grueling series of Congressional hearings cleared the brothers of conspiring to corner the market. Throughout the highly publicized proceedings, Bunker Hunt managed to maintain a dry sense of humor. Noting the press photographers who seemed to dog his every footstep, he observed: "At least I know they're using a little bit of silver every time they take my picture."

Although nations no longer use gold as a monetary yardstick, modern industrial civilization depends to a large degree in many other ways on all of the noble metals. As Sebastian Paul Musco, a dealer in precious metals, once said: "Because all industries are somewhat interrelated, there is scarcely one that would not be handicapped today if gold, silver, and the platinum metals were suddenly to become unavailable. A greatly surprised public would be similarly inconvenienced since these metals exert profound influences on its economy, health, safety, and convenience."

Jewelry, of course, remains the most familiar destiny of the precious metals. In increasingly affluent societies, the rings, pins and other articles that once belonged only to the rich and royal are now commonplace personal decorations. (The Japanese, for some reason, have long preferred platinum; in 1982, in fact, the Japanese jewelry industry accounted for fully 30 per cent of world platinum sales.)

Pure gold is too soft to withstand constant wear and is almost never used in jewelry. Rather, it is alloyed with varying amounts of other metals to give it strength and durability. The word "karat" is used to describe the amount of gold contained in an article. Pure gold is designated 24 karat; 18-karat gold, for example, contains 75 per cent gold and 25 per cent some other metal — usually silver or copper. Most modern gold jewelry is 14 karat, or 58 per cent pure gold. Various metals produce different colors when alloyed with gold; copper lends a pink or red hue, iron tints the metal blue and aluminum produces gold of a purplish cast.

Despite the noble metals' continuing popularity for use in jewelry, much greater quantities of gold, silver and platinum are put to other purposes. Medical and dental uses abound. Tasteless, nontoxic, resistant to acids and highly malleable as well, gold and silver can readily be shaped into dental caps, inlays and fillings that will withstand constant grinding and pressure. And gold also withstands the corrosive assaults of time: Archeologists have unearthed the remains of an Egyptian man whose perfectly preserved gold bridgework, installed 4,500 years ago, is the oldest known example of restorative dentistry.

While gold and its alloys are among the most effective metals for use in long-lasting dental reconstruction, the metal is not universally favored. In the Communist-bloc nations of Eastern Europe, for example, gold dental work is considered a show of imperialist decadence (as well as a waste of precious resources), and in cases where a capitalist dentist might emplace a gold cap or crown, his socialist counterpart will use stainless steel instead. But patients pay a price for such economy. Properly installed, a gold-based dental appliance will easily retain its size and shape for the life of the wearer; a less-malleable stainless-steel cap or crown, however, is exceedingly difficult to craft perfectly and may loosen after only a few years. In some Oriental cultures, on the other hand, a toothy display of dental gold is not viewed as decadent, but as a symbol of solid economic status. Some men in such societies have all their visible teeth encased in gold, and women flash glittery smiles highlighted by inset gold birds, hearts or flowers.

Physicians, too, use gold in their practices, though not nearly so visibly as do their dentist colleagues. The recognition that gold has certain curative powers goes back to ancient times. The Roman naturalist Pliny the Elder, who lived in the First Century of the Christian era, wrote of gold salves for the treatment of skin ulcers and the like. Medieval alchemists and medical practitioners included powdered gold in their potions for combating old age and advised their patients to drink from gold cups to prolong life. They also believed that gold mixed in drinks had the power to "comfort sore limbs," an obvious reference to the crippling effects of arthritis.

The use of gold in modern medicine began in 1890, when the pioneering German physician and bacteriologist Robert Koch revealed that gold cyanide and gold chloride — synthetic chemical compounds containing gold — inhibited growth of the bacillus that causes tuberculosis. Later researchers found that other gold compounds, usually administered by injection, were useful in the treatment of asthma and a variety of other conditions.

Antibiotics and other pharmaceuticals have supplanted gold compounds in most medical applications, but gold is still used, in combination with other chemicals, in the treatment of rheumatoid arthritis. The modern course of treatment, first applied in 1927 by a European physician, consists of injecting a gold compound into the body. While some individuals suffer toxic side effects, the treatments do produce moderate relief even in severe cases, though medical researchers have not yet determined precisely how the gold acts on bodily tissues to perform its therapeutic work.

Physicians also use gold in the X-ray radiation treatment of cancer, in-

How Silver Captures an Image

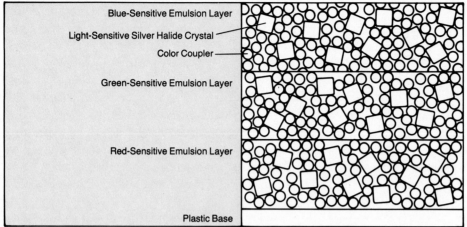

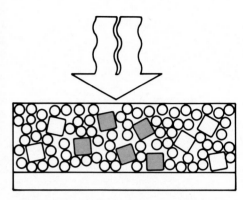

Almost a third of the silver mined in the world today is destined for use in photography; in the United States, fully 40 per cent of the silver consumed each year produces images on film and paper. The key to the process is the sensitivity of certain silver compounds, called silver halides, to light. How it works in color photography is detailed in the schematic drawings on these pages.

THE STRUCTURE OF COLOR FILM
This schematic cross section of color negative film shows its three layers of emulsion containing crystals of silver halides *(squares)*, surrounded by molecules of color couplers *(circles)*. The color couplers in each emulsion layer are sensitive to a different color of light — blue, green or red — and will create a different colored dye. The process is detailed for the red-sensitive layer at right.

EXPOSING THE LATENT IMAGE
When exposed to light, the silver halide crystal responds to the light energy, and tiny clusters of a few atoms of metallic silver form within the crystal. These particles of silver are so small they are not visible, even through a microscope. This invisible seed, and many others like it *(gray squares)*, form what is known as a latent image. The color couplers are not affected by exposure to light.

---

serting radioactive grains of the metal into malignant tumors, where the radiation destroys adjacent cancer cells. And tiny weights of gold, because they are heavy relative to their size and do not react with sensitive tissues, are sometimes implanted in the eyelids of certain patients to assist the working of diseased eye muscles.

Platinum, too, has many medical applications. Surgeons use scalpels and other instruments whose cutting edges retain their superb sharpness because of an extremely hard coating of platinum alloyed with iridium or ruthenium. (On a more workaday plane, shavers slice their whiskers away with keen-edged razor blades that have been similarly treated.) And platinum is sometimes alloyed with gold and used in restorative dentistry to give added strength.

Other medical applications of platinum are considerably more sophisticated. For example, platinum's electrical conductivity and unmatched compatibility with human tissues make the metal ideal for use in the tiny electrodes that are placed in the human body to deliver electrical impulses that stimulate muscles or brain cells. The most familiar such application is the electronic pacemaker, which ensures regular heart rhythms by delivering a steady series of electrical charges to pure platinum electrodes implanted in the heart muscle. Similar electrodes set deep within the brain and hooked up to a small electrical transmitter have been found to relieve patients suffering spinal-related pain that cannot be relieved by traditional narcotics. The mild electrical impulses stimulate the production of endorphins, the naturally produced chemicals that act to block pain.

Perhaps the most dramatic medical promise offered by platinum is in the treatment of cancer, though researchers do not know for certain how the

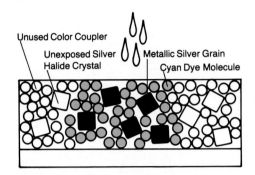

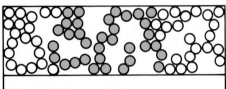

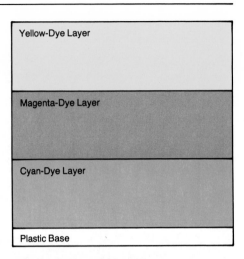

DEVELOPING THE IMAGE

Developing chemicals react with the crystals in which clusters of metallic silver have formed, converting the crystals entirely to grains of silver (*black squares*). The developer that has reacted with the exposed crystals then combines with the color couplers in the area surrounding the crystals, forming molecules of cyan dye (*circles*). The cyan dye blocks the transmission of red light through the negative during printing.

REMOVING THE SILVER

The film is next exposed to a solution that prevents any further reaction. Having guided the formation of the dye molecules which form the final image, the metallic silver grains and the unexposed silver halides are dissolved when the film is bleached and fixed. The silver washed out of the film in this step can be collected and recycled.

THE FINISHED NEGATIVE

In a completely processed color negative, all three of the color layers will contain an image. The layer of emulsion that is sensitive to blue light is dyed in yellow, the layer sensitive to green in magenta and the red-sensitive layer in cyan. The negative color image is then printed on photographic paper, which also has three emulsion layers; this step reverses the colors again, and the original hues of the subject appear in the print.

metal combats the disease. Powdered platinum compounds have been shown to be effective against leukemia in laboratory mice and rats, and to inhibit or completely eradicate some tumors in these animals. A platinum-based compound known as cis-Platinum II has been tested worldwide with encouraging results: In the United States, the National Cancer Institute used the compound along with other drugs in the treatment of testicular cancer in humans and reported a remission rate of 70 per cent. Promise has also been shown in the treatment of head and neck tumors with the platinum-based drug.

In addition to their medical applications, large amounts of the noble metals are employed in a variety of industrial and chemical processes. Silver is by far the most heavily used. A better conductor than copper of both heat and electricity, it finds wide application in computers, switches, thermostats and many other pieces of electrical equipment. The timer of a typical household dishwasher, for example, contains about 50 electrical contacts made of silver.

The largest consumer of silver is the photographic industry. Silver crystals, and their compounds with bromine, chlorine or iodine, are extremely sensitive to light and serve as the active ingredient in photographic paper and in most types of film (*pages 150-151*), including X-ray plates. Indeed, the greatest industrial user of silver in the world is the Eastman Kodak Company of Rochester, New York. In 1982, Kodak consumed about 25 per cent of the 119 million troy ounces of silver used in the United States.

Silver is also important in the printing industry, where it is used for photographic reproduction. Sizable amounts of the metal are also used in solders, brazing alloys and such traditional items as jewelry and tableware.

In a new application of silver's time-honored ability to reflect almost 100 per cent of the light that falls on it, a nine-acre field of mirrors traps the sun's energy to produce electricity. The array contains 222 heliostats — frames containing 25 mirrors each — that continually track the sun to focus its rays on a tank of fluid. Vapor from the heated fluid drives a generator.

Silver's post-World War II transition from a monetary role to widespread use as an industrial material has tended to make it less attractive to investors than gold and platinum, and it is increasingly viewed almost as a commodity — much like zinc, copper or lead. In fact, both silver and soybeans (which are a bellwether for general food prices) are important features on the rough-and-tumble commodities market of the Chicago Board of Trade; food prices are usually closely linked to inflation, and traders sometimes use silver as a hedge, buying the metal when soybean prices rise, and then selling it when soybeans begin to fall. Nevertheless, silver's low price — relative to gold and platinum — makes the metal available to large numbers of precious-metals investors, and it is widely purchased either through physical acquisition or paper transactions.

The international silver market is unique in the precious-metals field because the annual industrial consumption of the metal consistently exceeds the yearly output of the world's silver mines. This deficit in supply is made up in several ways. Governments and individuals hoard the metal and wait for the price to rise before putting it on the open market for sale. Another way of obtaining silver supplies is by recovering the silver in used X-ray plates and by melting down what is called scrap silver — jewelry, coins and tableware that have been discarded, and metal used in worn-out electronic equipment. About 150 to 200 million ounces of silver are generated annually in this fashion, as well as from recycled photographic film and developing chemicals. Eastman Kodak alone recovers 20 million ounces annually by recycling outdated film, black-and-white negatives and chemical solutions used to process film.

Gold, too, is widely used by science and industry, particularly in the electrical and electronics fields. A better conductor than any other metal except silver and copper, gold will not tarnish or oxidize and is thus prized for a number of applications in which its special wear-resistance and durability are worth the price. For example, the submarine telephone cables that snake across the ocean floor require repeaters, positioned every few miles along a cable's length, to amplify the strength of long-distance signals. These devices must be sturdy, operate reliably for long periods of time and resist the corrosive effects of their undersea environment; to achieve these goals, engineers specify that many of the repeater components be plated with gold.

Depending on how they are applied, thin films of gold can either absorb or reflect solar radiation. An almost transparent gold film will absorb and retain heat from the sun's rays, greatly enhancing a solar-energy collector's effectiveness. On the other hand, when the extensive plate-glass windows of large office buildings include a coating of transparent gold, most visible light is admitted but considerable high-energy infrared radiation is reflected, thus preventing overheating and reducing air-conditioning costs.

Yet for all the utility of gold and silver, the platinum metals offer the greatest boon to industry — and to agriculture as well. World food production would suffer grievously, and starvation might overtake parts of the world, were it not for the use of platinum in the production of synthetic nitrate fertilizers. Virtually all of the world's nitric acid, essential in the manufacture of these fertilizers, is made with a platinum-rhodium alloy catalyst woven into a fine gauze screen and heated. The gauze screen pro-

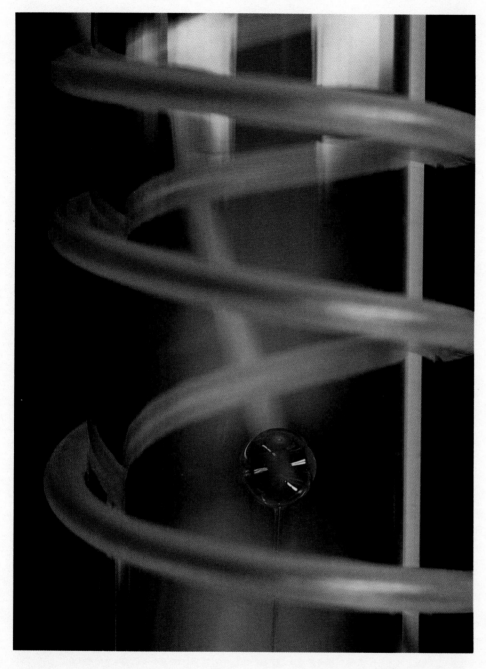

A glass pellet containing gaseous fuel for use in the still-experimental process of nuclear fusion gleams with a fresh coating of pure gold. A piece of gold foil placed in a glass cylinder with the pellet is heated until it vaporizes and condenses on the pellet; there, it will act to help control the speed of the fusion reaction.

duces at high temperatures the critical catalytic process for the oxidation of ammonia in the production of the necessary nitric acid *(pages 132-133)*.

The uses of platinum by industry are many and growing. Indeed, in the decades since World War II, more has been made of platinum metals than in all the years since their discovery in the early 1800s. Most significant, perhaps, is the employment of platinum as a catalyst in the chemical and petroleum industries, where it plays a key role in removing impurities during the refining process. Indeed, one estimate has it that fully 18 per cent of all manufactured goods rely in some way on materials that have been produced with platinum catalysts.

One of the newest and most rapidly growing applications is the automotive catalytic converter, which transforms carbon monoxide and other harmful wastes from gasoline engines into carbon dioxide and water. Since

1974, when the Environmental Protection Agency mandated installation of the devices on all new trucks and automobiles sold in the United States, millions upon millions of vehicles have been equipped with catalytic converters — to the point where, in 1982, American and Japanese automobile makers accounted for 25 per cent of world platinum consumption.

Because of platinum's high price — an ounce of the metal costs more than a like amount of gold — scientists are seeking a base-metal substitute for use in the converters. Until they succeed, serious consideration is being given to recovering platinum from cast-off converters in the same way that silver is recovered from used photographic materials. Though each converter contains only a very small amount of platinum, there are so many of them that one trade magazine observed: "Automobiles with catalytic converters are mines of precious metals on wheels."

To make the market tighter still, the use of the platinum metals in electronics and other technical fields has exploded since the 1950s. Platinum crucibles are used in the melting of optical glass for spectacles, camera lenses and television tubes. The manufacture of fiberglass for insulation and for fiberglass-reinforced products requires platinum-rhodium alloy plates with hundreds of tiny perforations through which molten glass is extruded in uniform threads. The process was first pioneered in the 1930s in the manufacture of such synthetic fibers as rayon; then as now, only platinum could stand up under the intense heat required by the operation. The development of fiber optics — the use of glass fibers as a replacement for copper wires in telecommunications equipment — depends on the high purity of the platinum equipment that produces the finely drawn threads of glass.

Palladium, one of the metals of the platinum group, is widely used in telephone-relay equipment, and with the development of electronic miniaturization, solid-state systems frequently require components of platinum and palladium or their alloys with gold or silver. Platinum's use has become so extensive in the electronics, chemical, petroleum, pharmaceutical, automobile and other industries that it has to be regarded in the 1980s as a critical industrial raw material.

In the future, platinum is likely to play a key role in the widespread development of fuel cells, which were used with success aboard the manned spacecraft in the U.S. space program. The fuel cell, invented in the early 19th Century, generates electricity by a reaction between oxygen and either hydrogen or a hydrocarbon such as acetylene or benzene. Large-scale fuel-cell plants capable of generating electricity in commercial quantities have been created with pilot projects in the cities of Tokyo and New York. Using the platinum electrodes as the critical catalytic agent in the fuel cell, such plants may well supplement existing generating units and act as stand-by emergency power stations.

Of all the modern uses of the noble metals, the most spectacular and far-reaching have been in the exploration of space. Platinum, gold and silver were critical in the evolution and success of the United States space program. Gold and silver electrical conductors are widely used in the high-technology components of all types of space vehicles. Extensive gold coating on the rocket engines' heat shields protects delicate instruments inside from damage by the searing temperatures during lift-off.

One unusual application of gold in the space program is as a lubricant. In

In the first use of gold in space, an astronaut maneuvering outside his spacecraft during a 1965 mission is protected from solar radiation by a gold film, only molecules thick, on his visor. Similar protection is afforded the slender "umbilical cord" linking him to the spacecraft.

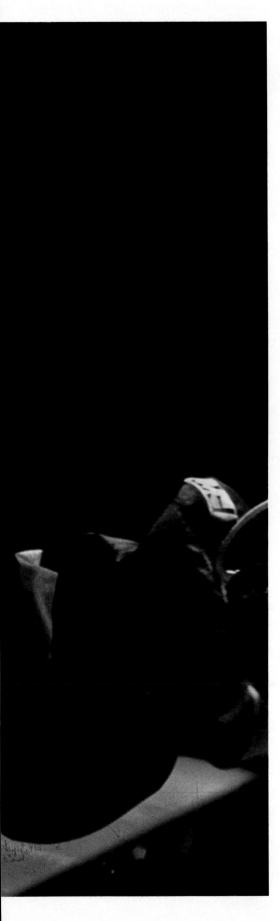

the vacuum of outer space, conventional oil-based lubricants can be affected by radiation, evaporation or chemical breakdown. Gold, on the other hand, is unaffected by such conditions, and in a finely powdered form, it is sometimes used as a solid lubricant for various mechanisms.

To the ancients, the noble metals were objects of wondrous beauty, imbued with the mystic qualities of distant heavenly bodies whose nature they seemed to mirror. For thousands of years, skilled artisans employed gold and silver — and later, platinum — to fashion exquisite jewelry and decorative pieces that in some cases are virtually all that remain today of once-flourishing civilizations. For thousands of years, too, men fought and died, or risked extreme privation, to add to their stores of these elusive, glittering minerals. Indeed, the history of nations is in a very real sense the history of the quest for precious metals, and the uses that have been made of them.

The modern world can also be judged by its uses of noble metals, though the standards of measurements have changed forever since Stone Age craftsmen shaped sacred amulets of gold, the metal of the sun. As the precious-metals dealer Sebastian Paul Musco has observed: "The extent to which a nation's economy utilizes precious metals such as platinum, palladium, gold and silver serves as a barometer of its technological proficiency." By such a standard, there can be no gainsaying the proficiency of a technological society that fuels its engines with petroleum products refined with platinum catalysts, that has built thriving electronics industries around switches and circuitry of gold and silver, and whose astronauts have peered out at the lunar landscape through helmet visors filmed with gold. And just as today's archeologists wonder at the intricate designs of ancient artifacts shaped of gold and silver, seeing in them the expressions of the highest aspirations of vanished cultures, so the archeologists of some future time may view such relics as a gold-coated electronic circuit or a platinum petroleum catalyst as the very essence of 20th Century civilization. **Ω**

THE MAJESTIC LEGACY OF GOLD

For 5,000 years, virtually every society on earth has expressed its vision of beauty in the shimmering splendor of gold. Life's necessities — tools and weaponry — had to be forged of harder stuff, but humanity has always esteemed gold for its loveliness, and for the power and prestige it conveyed.

Sumerians and Egyptians were probably the first to work with gold about 3000 B.C., having chanced upon the central discoveries about the metal: It melts at a comparatively low 1,000° F. and, if subjected alternately to heat and hammering, could be pounded paper thin, pulled into wire, formed into spheres, bent into chains, or even, when melted, poured into molds. The prehistoric techniques of working gold have changed little through the millennia.

Ancient peoples frequently attributed mystical and magical powers to gold. Egyptians wore gold amulets to ward off evil. West Africans prized golden fetishes. Both the gold-loving Mycenaeans and, much later, the Chimu of South America covered the faces of their dead princes with golden portrait masks. The Chinese believed that gold buried with the body prevented decay of the corpse.

This association of gold with death prompted many societies to stock the graves of their kings with hoards of treasure for use in the next life. The subsequent plunder of those tombs robbed the world of the vast majority of its gold artistry. The pieces that survive are but a sampling of mankind's golden heritage.

A gilded frog tops a wooden staff once carried by a 19th Century official in the West African kingdom of Asante. The wooden finial was first coated with a bonding substance, then covered with paper-thin gold leaf and heated, to melt the gold and bond it to the finial.

Delicately wrought tendrils of hair encircle a
ceremonial helmet worn by a Sumerian king
about 2500 B.C. To create this superb example
of the art of repoussé, artisans first pressed a thin
gold sheet into tacky pitch to hold the metal
securely, then embossed the design with blunt
chisels, working on what would become the
inside of the helmet. After the piece had been
removed from the pitch, cleaned and smoothed,
fine detail was added with chasing tools.

159

A Greek funeral wreath fashioned during the
Fourth Century B.C. shimmers with the
brilliance of hammered gold. The gold was
first heated, then hammered in a spiral pattern,
each stroke sliding across the piece in order
to avoid leaving hammer impressions, until the
metal became thin and pliable.

A skilled goldsmith from Chimu — a highly developed pre-Incan kingdom in Peru — shaped this golden beaker 900 years ago. The classic countenance took its shape from a carved wooden model placed inside the vessel. The thin metal was then hammered to conform to the model. This technique, called raising, has been in use for about 3,000 years.

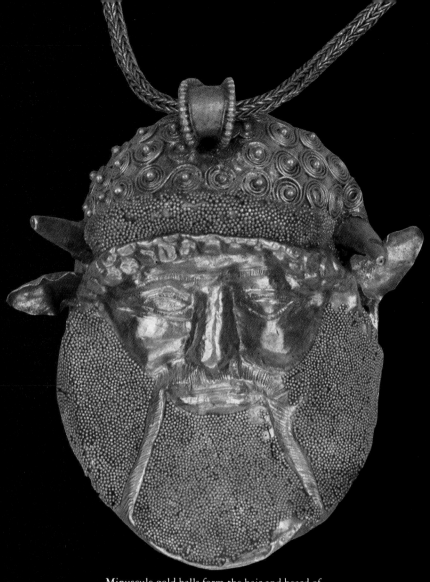

Minuscule gold balls form the hair and beard of the river god Achelous on this two-inch-high Etruscan pendant dating from the Sixth Century B.C. The spheres, as tiny as a few hundredths of an inch across, were soldered in place, using a technique called granulation. While no records exist of the methods used by the Etruscans, the small pellets were probably formed by letting molten gold splatter onto a smooth surface.

Golden stripes inlaid across the silver face of a warrior-goddess from the ancient kingdom of Thrace mirror the actual tattoos worn by high-ranking people of the time. The goddess adorns a greave, a piece of leg armor, found in the tomb of a prince who died about 375 B.C. To create inlay, a strip of gold is heated, then hammered into a precisely chiseled groove, the edges of which are beveled to grip the gold.

162

АПЛЪ ПАВЕЛЪ · МАРКО · АПЛЪ

Graceful figures of evangelists adorn the base of a gem-encrusted golden incense burner, a gift to the Church from the wife of the Russian Tsar Fëdor in the late 16th Century. In a process called niello, the design was etched into the gold, and a mixture of powdered silver, copper, lead and sulfur was melted into the fine incisions. The technique was little known in Europe at the time Russian goldsmiths mastered it, endowing their creations with a lush, velvety texture.

Emblazoned on a blue silk robe, a Chinese dragon embroidered with gold and silver thread bespeaks the power and wealth of the 18th Century Qing Dynasty emperor who wore it. Ancient peoples occasionally decorated garments with gold leaf, but when artisans learned to draw the metal into fine strands, they could produce more intricate and durable designs.

Set off by a luxuriant wreath of sable, this magnificent crown — believed to date back to the 11th Century Russian Tsar Monomakh — displays the sinuous intricacies of gold filigree. The art of filigree closely followed the discovery of a way to make gold wire by heating a thin strip of gold and pulling it through a tiny opening in drawplates. The wire is then bent into lacy patterns and soldered into place.

An extravagant saltcellar, on which Neptune
reigns over the salt in the basin below his trident
while Ceres, goddess of Earth, guards the
pepper in the temple to her right, is the only
surviving work in gold by the renowned
16th Century goldsmith Benvenuto Cellini.
The artist sculpted his masterpiece in clay,
covered it with a thin veneer of wax and an outer
layer of clay. He heated the piece to melt and
drain away the wax, then poured molten gold in
to replace the wax. Breaking away the outer
clay revealed the golden figures.

ACKNOWLEDGMENTS

For their help in the preparation of this book the editors wish to thank: **In Canada:** Ontario — International Nickel Company, Dennis Nagata; Ministry of Natural Resources, Thomas Patrick Mohide; **In Great Britain:** London — Timothy Green; Fiona Brown, Graham Williams, Consolidated Gold Fields PLC; Ian Cottington, Sarah Pleat, Johnson Matthey PLC; Tony Brewis, Mining Magazine; Patrick Finn, Spink and Son Ltd.; Charles Truman, Victoria and Albert Museum; Stoke-On-Trent — Robert Copeland, Spode Museum collection. **In South Africa:** Johannesburg — Public Relations Department, Anglo American Corporation of South Africa; Public Relations Department, Chamber of Mines; Public Relations Department, Johannesburg Consolidated Investment Company. **In the United States:** Alaska — (Fairbanks) Renée Blahuta; California — (Berkeley) Dr. and Mrs. Donald McLaughlin; University of California, The Bancroft Library-Regional Oral History Office; (Los Angeles) Don Chaput; (Santa Monica) Peter Keller, Gemological Institute of America; District of Columbia — Henry I. Dworshak, Shirley Kessel, J. Allen Overton, Kathryn Boggs

St. John, American Mining Congress; Margaret King, Brookings Institution; Dr. Mikolas Radvanyi, Winston Tabb, Library of Congress; Martin Luther King Memorial Branch, District of Columbia Library; David Pinkard, Mining Congress Journal; Walter W. Weinstein, National Bureau of Standards; Mary Smith, National Geographic Magazine; Roger Loebenstein, John M. Lucas, Bureau of Mines, Division of Non-Ferrous Metals, United States Department of Interior; United States Department of Interior Library; Florida — (Punta Gorda) Dr. Philip W. Guild; Illinois — (Chicago) Dr. Paul Brian Moore; Maryland — (Avondale) Dr. Stephen Cramer, Dr. David Flinn, Paul Moran, Richard Walters, Avondale Research Center, Bureau of Mines, United States Department of Interior; (Bethesda) Montgomery County Library; Massachusetts — (Woods Hole) David A. Ross, Woods Hole Oceanographic Institution; Nevada — (Carlin) R. S. Mattson, Carlin Gold Mining Company; New York — (Mamaroneck) Dr. Julius Weber; (New York City) Antoinette Fitapelli, Cartier, Inc.; David Graham, International Nickel Company Ltd.; International Precious

Metals Institute; Laura A. Rzasa, Thomas L. Richmond, Inc.; Pennsylvania — (Malvern) Joseph H. Povey, Johnson Matthey Inc.; (West Chester) David Lundy, Robert Staargaard, Johnson Matthey Inc.; Texas — (Houston) Sam Glorioso, National Aeronautics Space Administration; Virginia — (Reston) Dr. William Cannon, Michael Lee, Dr. Bruce R. Lipin, Dr. Wayne C. Shanks, United States Geological Survey; Washington — (Spokane) Raymond O. Hanson, Cathy Smith, Goodnews Platinum Company; Wisconsin — (Appleton) John Barlow, Earth Resources. **In West Germany:** Frankfurt — Dr. Walter Behne, Degussa; Karin Schütze, Christel Übelacker, Platin Gilde International; Hanau — Dr. Ralf Schrank, W. C. Heraeus; West Berlin — Dr. Roland Klemig, Heidi Klein, Bildarchiv Preussischer Kulturbesitz; Wolfgang Streubel, Ullstein Bilderdienst; Weissenstein — Manfred Kage, Astrid Kage.

Particularly useful sources of information and quotations used in this volume were from *A History of Platinum and its Allied Metals* by Donald McDonald and Leslie B. Hunt, London, Johnson Matthey, 1982.

The index was prepared by Gisela S. Knight.

BIBLIOGRAPHY

Books

Allen, Gina, *Gold!* Thomas Y. Crowell, 1964.

American Society for Metals, *Metals Handbook, Ninth Edition.* American Society for Metals, 1979.

Annual Silver Review and Outlook. J. Aron, July 1982.

Ballantyne, Verne H., *How and Where to Find Gold.* Arco Publishing, 1983.

Bateman, Alan M.:
Economic Mineral Deposits. John Wiley & Sons, no date.
The Formation of Mineral Deposits. John Wiley & Sons, 1951.

Baumann, Ludwig, *Introduction to Ore Deposits.* Scottish Academic Press, 1976.

Black, J. Anderson, *A History of Jewelry.* Park Lane, 1981.

Bray, Warwick, *Gold of El Dorado.* American Museum of Natural History/Harry N. Abrams, 1979.

Bronson, William, and T. H. Watkins, *Homestake.* Homestake Mining Company, 1977.

Cable, Mary, and the Editors of Tree Communications, *The African Kings.* Stonehenge Press, 1983.

Cabri, Louis J., ed., *Platinum-Group Elements: Mineralogy, Geology, Recovery.* Canadian Institute of Mining and Metallurgy, 1981.

Calder, Nigel, *The Restless Earth.* Penguin Books, 1972.

Carter, Howard, *The Tomb of Tutankhamen.* E. P. Dutton, 1954.

Cash, Joseph H., *Working the Homestake.* Iowa State University Press/Ames, 1973.

Casson, Lionel, *The Greek Conquerors.* Stonehenge Press, 1981.

Casson, Lionel, and the Editors of Time-Life Books, *Ancient Egypt.* Time-Life Books, 1965.

Chidsey, Donald Barr, *The California Gold Rush.* Crown Publishers, 1968.

Coleman, Robert G., *Ophiolites.* Heidelberg: Springer-Verlag, 1977.

Degens, Egon T., and David A. Ross, eds., *Hot Brines and Recent Heavy Metal Deposits in the Red Sea.* Springer-Verlag, 1969.

De Terra, Helmut, *Humboldt.* Alfred A. Knopf, 1955.

Dibner, Bern, *Agricola on Metals.* Burndy Library, 1958.

Drury, Wells, *An Editor on the Comstock Lode.* Farrar & Rinehart, 1936.

Durant, Will:
Caesar and Christ. Simon and Schuster, 1944.
The Life of Greece. Simon and Schuster, 1939.

Edey, Maitland A., and the Editors of Time-Life Books, *Lost World of the Aegean.* Time-Life Books, 1975.

The Editors of American Heritage, *The California Gold Rush.* American Heritage Publishing, 1961.

The Editors of Time-Life Books:
Light and Film, revised ed. Time-Life Books, 1981.
Volcano. Time-Life Books, 1982.

Edwards, I.E.S., *Tutankhamun: His Tomb and Its Treasures.* Metropolitan Museum of Art and Alfred A. Knopf, 1977.

Evans, Anthony M., *An Introduction to Ore Geology.* Blackwell Scientific Publications, 1980.

Fielder, Mildred, *The Treasure of Homestake Gold.* North Plains Press, 1970.

Fisher, Vardis, and Opal Laurel Holmes, *Gold Rushes and Mining Camps of the Early American West.* Caxton Printers, 1968.

Fitzhugh, Edward F., Jr., *Treasures in the Earth.* Caxton Printers, 1936.

Gibbon, Edward, *The Decline and Fall of the Roman Empire,* Vols. 1, 2 and 3. Modern Library, 1932.

Green, Timothy:
The New World of Gold. Walker, 1981.
The World of Gold Today. Walker, 1973.

Grunfeld, Frederic V., *The French Kings.* Stonehenge Press, 1982.

Hall, D.G.E., *A History of South-East Asia.* St. Martin's Press, 1968.

Hamblin, Dora Jane, and the Editors of Time-Life Books, *The Etruscans.* Time-Life Books, 1975.

Haywood, Richard Mansfield, *Ancient Rome.* David McKay, 1967.

Heezen, Bruce C., and Charles D. Hollister, *The Face of the Deep.* Oxford University Press, 1971.

Hepburn, A. Barton, *A History of Currency in the United States.* Augustus M. Kelley, 1967.

Hicks, Jim, and the Editors of Time-Life Books, *The Empire Builders.* Time-Life Books, 1974.

Hoberman, Gerald, *The Art of Coins and Their Photography.* Spink & Son with Harry N. Abrams, 1981.

Homestake Centennial 1876-1976. Homestake Mining Co., 1976.

Hoover, Herbert Clark, and Lou Henry Hoover, transls., *Georgius Agricola de re Metallica.* Dover Publications, 1950.

Horsfield, Brenda, and Peter Bennet Stone, *The Great Ocean Business.* Coward, McCann & Geoghegan, 1972.

Htin Aung, Maung, *A History of Burma.* Columbia University Press, 1967.

Humboldt, Alexander de, *Political Essay on the Kingdom of New Spain.* AMS Press, 1966.

Hurt, Harry, III, *Texas Rich.* W. W. Norton, 1981.

Idriess, Ion L., *Cyaniding for Gold.* Sydney: Angus and Robertson, 1939.

Jaffee, Bernard, *Crucibles: The Story of Chemistry.* Fawcett Publications, 1957.

Jensen, Mead L., and Alan M. Bateman, *Economic Mineral Deposits.* John Wiley & Sons, 1979.

Johnson, William Weber, and the Editors of Time-Life Books, *The Forty-Niners.* Time-Life Books, 1974.

Kemmerer, Edwin Walter, *Gold and the Gold Standard.* McGraw-Hill, 1944.

Kettell, Brian, *Gold.* Ballinger Publishing, 1982.

Knauth, Percy, and the Editors of Time-Life Books, *The Metalsmiths.* Time-Life Books, 1974.

Kramer, Samuel Noah, *The Sumerians.* University of Chicago Press, 1963.

Lamey, Carl A., *Metallic and Industrial Mineral Deposits.* McGraw-Hill, 1966.

Laurent, Antoine, *Oeuvres de Lavoisier,* Book 3. Paris: Imprimerie Impériale, 1865.

Leonard, Jonathan Norton, and the Editors of Time-Life Books, *Ancient America.* Time-Life Books, 1967.

Lewis, Robert S., *Elements of Mining*. John Wiley & Sons, 1964.

Linecar, Howard, *Coins and Coin Collecting*. Hamlyn, 1971.

Littlepage, John D., and Demaree Bess, *In Search of Soviet Gold*. Harcourt, Brace, 1938.

Lord, Eliot, *Comstock Mining and Miners*. Howell-North, 1959.

McDonald, Donald, and Leslie B. Hunt, *A History of Platinum and its Allied Metals*. London: Johnson Matthey, 1982.

McKinstry, Hugh Exton, *Mining Geology*. Prentice-Hall, 1948.

McPhee, John, *Coming into the Country*. Farrar, Straus and Giroux, 1977.

Mallowan, M.E.L., *Early Mesopotamia and Iran*. McGraw-Hill, 1965.

Marx, Jenifer, *The Magic of Gold*. Doubleday, 1978.

Mathewson, C. H., ed., *Modern Uses of Nonferrous Metals*. American Institute of Mining and Metallurgical Engineers, 1953.

Meyer, Jerome S., *The Elements, Builders of the Universe*. World Publishing, 1957.

Miller, Russell, and the Editors of Time-Life Books, *Continents in Collision*. Time-Life Books, 1983.

Mohide, Thomas Patrick:
Gold. Ontario: Ministry of Natural Resources, 1981.
Platinum Group Metals — Ontario and the World. Ontario: Ministry of Natural Resources, 1979.

Monaghan, Jay:
Australians and the Gold Rush. University of California Press, 1966.
Custer. Little, Brown, 1959.

Morrell, W. P., *The Gold Rushes*. London: Adam and Charles Black, 1940.

Neuburger, Albert, *The Technical Arts and Sciences of the Ancients*. London: Methuen, 1930.

Park, Charles F., Jr., and Roy A. MacDiarmid, *Ore Deposits*. W. H. Freeman, 1975.

Parr, J. Gordon, *Man, Metals and Modern Magic*. American Society for Metals and Iowa State College Press, 1958.

Pearce, Frank, *Last Call for HMS Edinburgh*. Atheneum, 1982.

Platinum/Palladium Review and Outlook. J. Aron, 1982.

Poss, John R., *Stones of Destiny*. Michigan Technological University, 1975.

Pryor, E. J., *Mineral Processing*. London: Elsevier Publishing, 1965.

Rapson, W. S., and T. Groenewald, *Gold Usage*. Academic Press, 1978.

Rawlings, G. B., *Ancient/Medieval/Modern Coins and How to Know Them*. Ammon Press, 1966.

Rhodes, Frank H. T., *Geology*. Golden Press, 1972.

Rickard, T. A.:
A History of American Mining. McGraw-Hill, 1932.
Man and Metals. Arno Press, 1974.
The Romance of Mining. Toronto: Macmillan Company of Canada, 1944.

Riley, Charles M., *Our Mineral Resources*. John Wiley & Sons, 1959.

Robbins, Peter, and Douglass Lee, *Guide to Precious Metals and their Markets*. Van Nostrand Reinhold, 1979.

Rolfe, Sidney E., *Gold and World Power*. Harper & Row, 1966.

Rose, T. Kirke, *The Precious Metals*. London: Archibald Constable, 1909.

Satterfield, Charles N., *Heterogeneous Catalysis In Practice*. McGraw-Hill, 1980.

Savin, I. V., *Physical Metallurgy of Platinum Metals*. Moscow: MIR Publishers, 1978.

Sawkins, Frederick J., et al., *The Evolving Earth*. Macmillan, 1974.

Scientific American, Continents Adrift and Continents Aground. W. H. Freeman, 1976.

The Search for Alexander. Greek Ministry of Culture and Sciences and New York Graphic Society, 1980.

Seidman, Laurence I., *The Fools of '49: The California Gold Rush, 1848-1856*. Alfred A. Knopf, 1976.

Singer, Charles, E. J. Holmyard and A. R. Hall, eds., *A History of Technology*, Vols. 1 and 2. Oxford University Press, 1954.

Sinkankas, John, *Mineralogy*. Van Nostrand Reinhold, 1964.

Smith, Ernest A., *The Platinum Metals*. London: Sir Isaac Pitman & Sons, 1925.

Smith, Grant H., *The History of the Comstock Lode 1850-1920*. Nevada State Bureau of Mines and McKay School of Mines, 1974.

Stokes, William Lee, and Sheldon Judson, *Introduction to Geology*. Prentice-Hall, 1968.

Sutherland, C.H.V., *Gold: Its Beauty, Power and Allure*. McGraw-Hill, 1969.

Swanberg, W. A., *Citizen Hearst*. Charles Scribner's Sons, 1961.

Tebbel, John, *The Life and Good Times of William Randolph Hearst*. E. P. Dutton, 1952.

Thompson, James Westfall, *Economic and Social History of the Middle Ages (300-1300)*. Frederick Ungar, 1966.

Toutain, Jules, *The Economic Life of the Ancient World*. Barnes & Noble, 1951.

Toynbee, Arnold J., *A Study of History*. Oxford University Press, 1946-1957.

Trigger, Bruce G., *Nubia under the Pharaohs*. London: Thames and Hudson, 1976.

Trustees of the British Museum, *The Times* and *Sunday Times, Treasures of Tutankhamun*. London: British Museum, 1972.

Twain, Mark, *Roughing It*. American Publishing, 1872.

Vines, R. F., *The Platinum Metals and Their Alloys*. International Nickel Company, 1941.

Walker, Bryce, and the Editors of Time-Life Books, *Earthquake*. Time-Life Books, 1982.

Walker, Robert Tunstall, and Woodville Joseph Walker, *The Origin and Nature of Ore Deposits*. Walker Associates, 1956.

Wallace, Robert, and the Editors of Time-Life Books, *The Miners*. Time-Life Books, 1976.

Watkins, T. H., *Gold and Silver in the West*. American West Publishing Company, 1971.

Wear, Ted G., *Ancient Coins: How to Collect for Fun and Profit*. Doubleday, 1965.

White, Benjamin, *Silver: Its History and Romance*. Tower Books, 1971.

Wilson, Neill C., *Silver Stampede*. Macmillan, 1937.

Winkler, John K., *William Randolph Hearst: A New Appraisal*. Hastings House, 1955.

Woolley, Sir Leonard:
The Art of the Middle East. Crown Publishers, 1961.
Excavations at Ur. London: Ernest Benn, 1954.
History Unearthed. London: Ernest Benn, 1963.

Young, George J., *Elements of Mining*. McGraw-Hill, 1946.

Young, Otis E., Jr., *Western Mining*. University of Oklahoma Press, 1970.

Periodicals

Anderson, James A., "Gold: Its History and Role in the U.S. Economy and the U.S. Exploration Program of the Homestake Mining Co." *Mining Congress Journal*, January 1982.

"Big Boom in a Barbarous Relic." *Time*, February 26, 1979.

Boraiko, Allen A., "A Mineral of Excellent Nature: Silver." *National Geographic*, September 1981.

Clark, Eugenie, "The Strangest Sea." *National Geographic*, September 1975.

Davison, Lonnelle, "Platinum in the World's Work." *National Geographic*, September 1937.

Filstrup, Jane Merrill, "Precious Platinum." *Town & Country*, June 1983.

"The Great Sell-Off." *Time*, January 14, 1980.

Guzzardi, Walter, Jr., "Gold: The Huge Find in Roy Ash's Backyard." *Fortune*, December 27, 1982.

Heywood, A. E., "The Recovery of Platinum from Ammonia Oxidation Catalysts." *Platinum Metals Review*, January 1982.

"A Hunt Crony Tells All." *Fortune*, June 30, 1980.

Hunt, L. B., "The First Real Melting of Platinum." *Platinum Metals Review*, April 1982.

Hunt, L. B., "Platinum in the Decoration of Porcelain and Pottery." *Platinum Metals Review*, October 1978.

Jackson, Dan, "Carlin Gold." *Engineering and Mining Journal*, July 1983.

Johnston, Charles, "Platinum Mining in Alaska." *Platinum Metals Review*, April 1962.

Judge, Joseph, "Greece's Brilliant Bronze Age." *National Geographic*, February 1978.

Kampf, Anthony R., and Peter C. Keller, "The Colorado Quartz Mine." *The Mineralogical Record*, November/December 1982.

Leicht, Wayne, "California Gold." *The Mineralogical Record*, November/December 1982.

McQuiston, Frank W., Jr. and Robert W. Hernlund, "Newmont's Carlin." *Mining Congress Journal*, no date.

Matthews, Samuel W., "This Changing Earth." *National Geographic*, January 1973.

Nesbit, Wilbur D., "Worth More than Its Weight in Gold." *Harper's Weekly*, June 22, 1912.

"The Platinum Situation." *The Literary Digest*, August 17, 1918.

"Recovering Platinum from Auto Catalysts." *Chemical Week*, March 23, 1983.

Renfrew, Colin, "Bulgaria's Ancient Treasures." *National Geographic*, July 1980.

Rosenberg, Barnett, "Some Biological Effects of Platinum Compounds." *Platinum Metals Review*, April 1971.

St. John, Jeffrey, "Reflections on Glittering Generations." *Mining Congress Journal*, December 1982.

Scott, David A., and Warwick Bray, "Ancient Platinum Technology in South America." *Platinum Metals Review*, October 1980.

"Stampede for Precious Metals." *Time*, January 28, 1980.

"A Talkfest with the Hunts." *Fortune*, August 11, 1980.

"The Treasure of Serra Pelada." *Life*, December 1980.

"The Treasure of Serra Pelada." *Time*, September 1, 1980.

White, Peter T., "The Eternal Treasure: Gold." *National Geographic*, January 1974.

Williams, Colin J., Herman J. Seidmann Jr. and Robert E. Hawley, "Is Fire Assay Here to Stay?" *American Laboratory*, August 1975.

Wilson, Wendell E., "The Gold-Containing Minerals: A Review." *The Mineralogical Record*, November/December, 1982.

Young, Gordon, "The Miracle Metal Platinum." *National Geographic*, November 1983.

Other Publications

Adair, William, *The Frame in America, 1700-1900: A Survey of Fabrication Techniques and Styles*. The American Institute of Architects Foundation, 1983.

Armstrong, Frank C., ed., *Genesis of Uranium- and Gold-Bearing Precambrian Quartz-Pebble Conglomerates*. U.S. Government Printing Office, 1981.

Brobst, Donald A., and Walden P. Pratt, eds., *United States Mineral Resources*. U.S. Government Printing Office, 1973.

Encyclopaedia Britannica, Vol. 10. William Benton, 1960.

Gold in the Monetary System — A Re-Examination. International Gold Corporation, 1981.

Gold Mining and Recovery: Yesterday and Today. Gold Information Center, no date.

Gries, John Paul, *Providing New Sources of Mineral Supply*. Bureau of Mines, U.S. Department of the Interior, 1979.

Heady, H. H., and K. G. Broadhead, *Assaying Ores, Concentrates, and Bullion*. Bureau of Mines, U.S. Department of the Interior, 1977.

McLaughlin, Donald H., *Careers in Mining Geology and Management, University Governance and Teaching*. Regents of the University of California, 1975.

Mertie, John B., Jr., *Economic Geology of the Platinum Metals*. U.S. Government Printing Office, 1969.

Musco, Sebastian Paul, "Economics of Refining of Precious Metals." *Economic Aspects of Precious Metals, IPMI,* February 7-8, 1979.

The National Research Council, "Supply and Use Patterns for the Platinum-Group Metals." National Academy of Sciences, 1980.

Petrascheck, W. E., *Metallogenetic and Geochemical Provinces*. Springer-Verlag, 1972.

Platinum-Group Metals. Bureau of Mines, U.S. Department of the Interior, 1983.

"Price Volatility in the Silver Futures Market." *Hearings before the Subcommittee on Agricultural Research and General Legislation, Committee on Agriculture, Nutrition and Forestry, May 1 and 2, 1980*. U.S. Government Printing Office, 1980.

Retrospective Louis Cartier. Cartier Inc., 1982.

Roberts, Ralph J., "Alinement of Mining Districts in North-Central Nevada." *Short Papers in the Geological Sciences*. U.S. Government Printing Office, 1960.

Schwartz, Anna J., *The Role of Gold in U.S. Experience, 1934-1981*. National Bureau of Economic Research, no date.

Treasures of Tutankhamun. British Museum, 1972.

Treasures of Tutankhamun. Metropolitan Museum of Art, 1976.

PICTURE CREDITS

The sources for the illustrations that appear in this book are listed below. Credits from left to right are separated by semicolons, from top to bottom by dashes.

Cover: Dane Penland, courtesy Smithsonian Institution. 6, 7: Harold and Erica Van Pelt © 1982, courtesy Kristalle Collection; Harold and Erica Van Pelt © 1982, courtesy Bill Larson Collection. 8, 9: Harold and Erica Van Pelt © 1982, courtesy Kristalle Collection. 10, 11: Nelly Bariand, Paris; Harold and Erica Van Pelt © 1983, courtesy Kristalle Collection. 12, 13: Wendell E. Wilson; Harold and Erica Van Pelt © 1982. 14, 15: Wendell E. Wilson. 16: Photoresources, Dover, England. 19: Hirmer Fotoarchiv, Munich. 20: Map by Bill Hezlep. 21: Ozan Sagdic, courtesy Ankara Archaeological Museum. 22: Art by Bill Hennessy. 24, 25: Art by John Drummond; Egyptian Expedition, The Metropolitan Museum of Art, except bottom right, Ullstein Bilderdienst, Berlin (West). 26, 27: © Lee Boltin. 29: D. Widmer, Basel, Switzerland. 31: Courtesy Germanisches Nationalmuseum, Nuremberg. 32: © Jonathan T. Wright. 34, 35: M. Ricciardi from Gamma-Liaison. 36: M. Ricciardi from Gamma-Liaison (2) except top right, Claus Meyer © 1980 from Black Star. 38: © Lee Boltin; drawings by Walter Hilmers. 42-46: © Gerald Hoberman, courtesy Spink and Son Ltd., London. 47, 48: Gerald Hoberman, Cape Town, South Africa. 49: © Gerald Hoberman, courtesy Spink and Son Ltd., London. 50: © Manfred Kage from Peter Arnold, Inc. 53: Harold and Erica Van Pelt © 1982, courtesy Bill Larson Collection. 55-57: Art by Walter Hilmers. 59: © Olaf Medenbach, Witten, Germany; Harold and Erica Van Pelt, courtesy F. John Barlow — Harold and Erica Van Pelt, courtesy F. John Barlow; © Olaf Medenbach, Witten, Germany (3). 61: Map by Bill Hezlep. 62, 63: Woods Hole Oceanographic Institution. 66, 67: Art by Greg Harlin from Stansbury, Ronsaville, Wood, Inc., except left, U.S. Geological Survey. 68, 69: Dr. Julius Weber. 71: Dr. David A. Ross from Woods Hole Oceanographic Institution; art by Greg Harlin from Stansbury, Ronsaville, Wood, Inc. (3). 72: Fred Ward from Black Star. 74-83: Art by Ken Townsend. 84: Courtesy History Division Los Angeles County Museum of Natural History. 86: © J-C Pinheira from TOP, Paris — Christopher Springmann © 1981 from The Stock Market — © Dewitt Jones © 1980 from Woodfin Camp Inc. 87: Courtesy Consolidated Gold Fields PLC, London. 90: California State Library, California Section Photograph Collection. 91: Art by Maria T. Estevez. 92, 93: Courtesy The Bancroft Library. 94, 95: Courtesy Museum of New Mexico. 96: Henry Beaufoy Merlin, Mitchell Library, Sydney, courtesy Australian Information Service, London. 98, 99: Courtesy The Bancroft Library. 101: © L. B. Hollister, courtesy Homestake Mining Company. 102, 103: Art by Bill Hennessy. 104, 105: Luisia Ricciarini, Milan. 106, 107: Dieter Blum from Peter Arnold, Inc. 108, 109: Courtesy Anglo American Corporation of South Africa; courtesy Consolidated Gold Fields PLC, London. 110, 111: James L. Stanfield, © National Geographic Society. 112, 113: Courtesy Anglo American Corporation of South Africa; Harry Redl © 1980 from Black Star. 114, 115: Stockphotos, Inc., inset Marka, Milan. 116: © Rainer Bode, Bochum, Germany. 119: Platin Gilde International, Frankfurt, courtesy Museum fur Völkerkunde, Staatliche Museen Preussischer Kulturbesitz, Berlin (West). 121: From Oeuvres de Lavoisier, Publiées par les soins de l'Instruction Publique, Paris 1865. 124: Cartier, Inc. — Victoria and Albert Museum, Crown Copyright, London (2). 125: Spode Museum Collection, Stoke-on-Trent — James L. Amos, photographed at The Metropolitan Museum of Art, Purchase, 1974. Gift of Dr. and Mrs. A. L. Garbat, Manya Garbat Starr, and Julian A. Garbat, by exchange, and Harris Brisbane Dick Fund, by exchange. (1974.164). © National Geographic Society. 128, 129: R. A. Hanson Co. 131: Johnson Matthey Inc. 132: © Manfred Kage from Peter Arnold Inc. 133: Johnson Matthey Metals, Ltd. — © Manfred Kage from Peter Arnold Inc. 134: DVA/Hamann, Stuttgart. 135: Munroe Studios, courtesy F. John Barlow. 136-138: © Manfred Kage from Peter Arnold Inc. 142: Fred Ward © 1978 from Black Star. 145-147: Fred Ward © 1981 from Black Star. 150, 151: Art by Frederic F. Bigio from B-C Graphics. 152: Fred Ward © 1981 from Black Star. 154: © Dan McCoy from Rainbow. 156, 157: NASA. 158, 159: Doran Ross; Iraq Museum, Iraq, courtesy The Bridgeman Art Library, London. 160: Spyros Tsavdaroglou, Athens. 161: Lee Boltin. 162, 163: Réunion des Musées Nationaux, Paris; Erich Lessing from Magnum. 164, 165: Boris Kuznetsov/VAAP, Moscow. 166: B. Bogdanova/VAAP, Moscow. 167: The Metropolitan Museum of Art, Gift of Lewis Einstein, 1954. 168, 169: Erich Lessing, courtesy Kunsthistorisches Museum, Vienna.

INDEX

Monomakh, Tsar, crown of, *166*
Montana: gold mines, 99; Stillwater platinum deposits, 80
Morse, Samuel, 128
Mother Lode, California, 6, 64, 76
Mountain building, 60, *78-79,* 80, *81*
Muckers, 101, *109*
Murmansk Run, 139
Musco, Sebastian Paul, 148, 157
Mycenaeans, 158; burial mask, *16*

N

Nevada: Comstock Lode, 85, 86, 96-97, *98-99;* Tuscarora gold find, 106-107
New Jersey, platinum refining plant, *131*
Newmont Mining Corporation, 107
New World: discovery of, 33, 37; lure of gold, 33, 37, 39, 118; mining, 39, 40, 89-90, 118-119
Nickel, 80, 130-131
Niello, gold artifact, *164-165*
Nitric acid, 120, 122; production, *132-133,* 134-135, 153-154
Noble metals: vs. base metals, 18; melting points of, *table* 120, 123, 126; qualities of, 18, 54, 55, *diagrams* 56-57, 58, 117
Nome, Alaska, gold beach placer near, 65
North America: British and French search for gold, 39; western edge, precious metals of, 64, 65, 75
North Carolina, gold strike of 1799, 40
Nubia, gold mines of, 19
Nubian Desert, 86
Nuggets, 82; of gold, *14-15,* 18, 35, *36,* 65, 68, *86,* 90, 107; of gold, hammered to gold leaf, *22;* gold, largest, *96;* of platinum, 79, *116, 118,* 119

O

Obduction, 78-79
Ocean floor: deposits of precious metals on, 72, 73, *74-75;* mining, 73, 74; search for precious metals on, 72, 73; spreading, 70, *diagram* 71, 72, 73
Oceanic crust, 76-79; forming of new, 60, *diagram* 71, 73, *74-75*
Oceanic plates, 64, 72, 74, 76-79
Oceans, 60; gold in water of, 70
Ontario gold mine, Utah, 99
Open-pit mines, gold, 86, 100, 107
Ophiolite suite, 78-79; weathered, 78, 79
Ophir, 96, 97
Ore deposits, 58; age of, 61; formation of, 52, 60, 64; global distribution near plate boundaries, 60, *map* 61, 64, 73; hydrothermal, 60, 64, 70, *diagram* 71, 73, *74-77;* in obduction, 78-79; placers, 64-65, *66-67, 68-69,* 78, 79, 80, *82-83;* primary, *80-81,* 82; secondary, 80, 82. See also Gold deposits; Platinum deposits; Silver deposits
O'Riley, Peter, 96
Oroville, California, 90
Osmium, 126, 131; melting point of, *table* 120, 126
Ostwald, Wilhelm, 134
Outcroppings, 80-*81,* 96

P

Pacemaker, electronic, 150
Pacific Plate, 60, *map* 61, 64
Palladium, 123, 127, 130, 132, 155; alloys, 123, 126, 127; crystals, *136-137;* melting point of, *table* 120, 123; refining, *131*
Panning, *36, 84,* 86, 90-91; pan and rocker, 90, *91*
Paper money, 40, 141-142; bimetallic or silver standard, 41; gold standard, 41, 141, 143
Patio process, 89
Paul, Grand Duke, of Russia *(later* Tsar), 122
Penny, English 10th Century, *49*
Penrod, Emanuel, 96
Persia, ancient, 23
Peru, 33, 37, 104; Chimu metalworking, *38,* 158, *161;* Spanish subjugation of, 37, 39
Petroleum industry, platinum metals used in, 118, 126, 154, 155, 157

Philip II, King of Macedonia, 23, 29
Photography: color, silver halides in, *diagrams* 150-151; platinum metals used in, 134, 135; silver used in, 134, 141, 145, 151, 153
Pillow lava, *62-63, 74*
Pizarro, Francisco, 37, 39
Placer deposits, 64-65, *66-67,* 68, 78, 79, 80, *82-83;* alluvial (stream), 64-65, *66-67,* 68-69, 79, *82-83;* beach, 65, 79, *83;* biochemical, *66,* 67, 69; eluvial, 64, *82;* eolian, 64; mechanical, *66,* 67; prospecting and mining of, 86, 87, *90-91,* 94, 96, 107
Placerville, California, 90
Planets, role in ancients' views of metals, 51-52
Plate tectonics, 60, 61, 64, 70, *diagram* 71, 72-73, 74, *76-81,* 86
Platina, 118
Platinum, 18; appearance of, 54, 118; ascendancy of, 117-118; atomic bonding of, 120, 122; atomic structure of, 55; capable of welding, 120-121; as catalyst, 117-118, *132-133,* 134-135, 153-154, 155; chemical inertness of, 54, 64, 117, 135; colored precipitates of, 120, 123, 126; as conductor of electricity, *diagram* 56, 150; in counterfeit coins, 120; density, 54; in dentistry, 135, 150; ductility, 117; durability, 18; gold compared to, 117, 120, 122; hardness, 54; high melting point, 117, *table* 120, 121-122; industrial uses and value, 117-118, 126-127, 131, *132-133,* 134-135, 141, 153-155; luster, 54, *diagram* 57; malleability, 120, 121, 122-123, 126; medical applications of, 126, 150-151; modern uses of, 135; price of, 117, 124, 154; rareness (percentage of earth's crust), 54, 117; refining, 122-123, 124, 128, 130-*131;* Russian monopoly of past, 127, 128; slow acceptance of, 118-122; in space-age technology, 155; Spanish monopoly of past, 122-123, 127; tensile strength, 117, 122
Platinum alloys, 128, 134, 155; with iridium, 126, 150; with palladium, 126; with rhodium, 132, 153; with ruthenium, 126, 150
Platinum artifacts, 117, 118, *119,* 122, 123, *124-125;* ancient, 118; jewelry, 54, 117, 122, *124,* 126, 135, 148
Platinum deposits, 54, 58, 117; alluvial placers, 65, 79, 134; beach placers, 65, 79, *128-129,* 134; eluvial placers, 64; found in ophiolite suite, 78-79; global distribution of, *map* 61; in Goodnews Bay, Alaska, *128-129,* 134; grains, 117, 118, 134; nuggets, 79, *116,* 118, 119; on ocean floor, *74-75;* outcroppings, 80-*81;* placers, 64, 65, 78, 79, 80, 82, 130, 134; primary, *80-81;* in Russia, 127-128; in South Africa, 68, 130-131; in South America, 118, 119, 127; in sulfides, *81;* underground veins, 130-131
Platinum group of metals, 80, 117, 118, 123, 126-127; alloys, 123, 126, 128; melting points, *table* 120; uses of, 123, 126-127, 128, 130, 153-155
Platinum mining, 118; extraction from ore, 58, 130-131; placer mining, *128-129,* 134; underground, 130-131
Pliny the Elder, 149
Pneumatic drill, 100, 109, *110-111*
Poor man's ore, 82, 107
Porcelain, Dresden, 30; platinum-plated, 124, *125;* silver plating on, 124
Pre-Columbian artifacts, 37, 39, 118, *119,* 158, *161*
Priestley, Joseph, 121
Primary deposits, *80-81,* 82
Prospecting, 65, 82, *84,* 85-86, *90-91,* 96, 99, 107
Proustite, 59
Pyramids: Babylon, 19; Egypt, 20, 24
Pyrargyrite, 59
Pyrite, *53,* 58
Pythagoras, 30

Q

Qing Dynasty, gold and silver embroidery, *167*
Qualities of noble metals, 18, 54, 55, *diagrams* 56-57, 58

Quartz, 65, 69; host rock of gold, *11,* 86, 87, 90, 96, 99, 130
Quartzite layers, Witwatersrand, 67, 69
Quibdó, Colombia, 119
Quicksilver, 51. See also Mercury
Quimbaya Indians, 38

R

Raising, gold-working technique, *161*
Rammelsberg mine, Harz, Germany, 88
Rand, the. See Witwatersrand
Rangoon, Burma, Buddhist temple, *32*
Rareness, of precious metals, 54
Rayon, 134, 155
Red Sea rift, 70, *diagram* 71, 72, 73; mineral wealth at, 70, *71,* 73, 86; sediment core samples, *71,* 73
Reflectivity, 54, *diagram* 57, 123, *152,* 153
Religious uses and meanings, of gold, *19-21,* 33, 37, 158
Repoussé gold working, *158-159*
Rhodium, 123, 126, 127, 130, 131, 132, 153, 155; melting point of, *table* 120, 123
Rich man's gold, 96, 107
Rickard, Thomas A., 41
Rift zones, 70, *diagram* 71, 74
Ríotinto, Spain, silver mines of, 88
Roberts: Ralph J., 106
Rock: erosion, 64, 79, *81, 82;* types of, 58. See also Hard-rock mining
Rocker, 90, *91,* 94
Rome, ancient, 20, 23, 28-29, 30, 47, 73, 88; aureus, 23, 28, *46*
Roosevelt, Franklin D., 142
Royal Academy of Science (France), 121, 122
Royal Society (of London), 119
Russia: filigree gold artifact, 11th Century, *166;* niello gold artifact, 16th Century, *164-165;* platinum monopoly of last century, 127, 128. See also Soviet Union
Russian Plate, 78
Rustenburg, South Africa, 130, 131
Ruthenium, 126, 130, 131, 150; melting point of, *table* 120, 126

S

Sahara, 86
Saxony, mining tradition of, 88-89, 97
Schliemann, Heinrich, 21
Scrap refining, for silver reclamation, *145-147,* 153
Sea-floor spreading, theory of, 70, *diagram* 71, 72, 73
Sea water, gold in, 70
Secondary deposits, 80, 82. See also Placer deposits
Sedimentary core samples, Red Sea, *71,* 73
Sedimentary rock, 58, 60, 67, 69, *diagram* 71, 73
Separation methods, 58, 86, 87, 88, 89; electrolysis, 130; flotation, 131; panning, *36, 84,* 86, 90-91; sluicing, *36,* 86, *91,* 94, 107; smelting, 87, 130. *See also* Chemical separation methods; Heating; Long tom; Rocker
Serra Pelada, Brazil, gold rush of 1980-1983, *34-36,* 107
Shwe Dagon pagoda, Rangoon, *32*
Siberia, platinum deposits of, 127
Siberian Plate, 78
Sierra Nevada, 60, 70, 90, 94, 96
Silicosis, 100
Silver: ancient concepts of, 51-52; arborescent, *12-13, 72;* atomic structure of, 55; as conductor of electricity, 54, *diagram* 56, 151; as conductor of heat, 54, 57, 151; crystalline nature of, 6; crystals, 6-7, *72;* density, 54; in depletion gilding, *38;* durability, 18, 42; in history, 21, 23, 40-41; industrial uses of, 141, 151, *152,* 153; luster, 54, *diagram* 57; malleability, 54; as medium of trade, 21, 23, 28-29, 42, 141, 153; melting point, *table* 120; photographic uses of, 134, 141, 145, 151, 153; price of, 145, 148, 153; rareness (percentage of earth's crust), 54; relative chemical inertness of, 54; in space-age technology, 155; storage at West

Time-Life Books Inc. offers a wide range of fine recordings, including a Rock 'n' Roll Era series. For subscription information, call 1-800-621-7026 or write Time-Life Music, P.O. Box C-32068, Richmond, Virginia 23261-2068.